Emerging Micro - and Nanotechnologies

Edited by Ruby Srivastava

Published in London, United Kingdom

IntechOpen

Supporting open minds since 2005

Emerging Micro - and Nanotechnologies
http://dx.doi.org/10.5772/intechopen.77527
Edited by Ruby Srivastava

Contributors
Yuchao Li, Zhiyong Gong, Selvakumar Varadarajan.Subramani, Rashmi Aradhya, Madhu B.M., Poornima Venkataramanaiah, Javier Martinez, Steven Voldman, Klaudia Freire

First published in London, United Kingdom, 2020 by IntechOpen
IntechOpen is the global imprint of INTECHOPEN LIMITED, registered in England and Wales, registration number: 11086078, 7th floor, 10 Lower Thames Street, London, EC3R 6AF, United Kingdom
Printed in Croatia

British Library Cataloguing-in-Publication Data
A catalogue record for this book is available from the British Library

Additional hard and PDF copies can be obtained from orders@intechopen.com

Emerging Micro - and Nanotechnologies
Edited by Ruby Srivastava
p. cm.
Print ISBN 978-1-78985-597-5
Online ISBN 978-1-78985-598-2
eBook (PDF) ISBN 978-1-78985-100-7

Meet the editor

Dr Ruby Srivastava, a theoretical physicist, is the principal investigator on her third consecutive DST (SR/WOS-A/CS-69/2018) project, with CSIR-CCMB, Hyderabad. After fifteen years in teaching, she started her research career and published several solo-authored papers, review articles, and seven book chapters. Awarded with Quarterly Franklin Membership, she is also member of the Cambridge Scholar Publishing and editor of many book projects. One of her books was selected to be in the Web of Science™ Core Collection. Her publication (*Phys. Chem. Chem. Phys.*, 2014) is rated #1 and two of her publications have been selected for Longuet-Higgins Early Career Researcher Prizes. She is a reviewer for many of the leading journals and is rated at 0.01% of the most read authors by Academia.edu.

Contents

Preface

Welcome to the emerging field of micro and nanoelectronics!

The new generation of nanotechnological devices presents increasingly formidable multidisciplinary challenges at the most fundamental level, which result in an urgent need for long-term research based on a scientific approach. These technologies are forcing changes to the way we view and use communications and information. The new technologically innovative devices need understanding of underlying physical mechanisms, and removal of present technological limitations. With a strong emphasis on experimentation and fabrication, the applications span the areas of nano-devices, optical communications, sensors, and biomedical devices. This book is all about the optoelectronic and biomedical applications of micro and nanotechnologies, in which authors from all over the world have contributed their research.

Chapter 1 deals with the issues of electrostatic discharge (ESD), electrical overstress (EOS), and latchup for VLSI microelectronics with a discussion on physical models, failure mechanisms, and design solutions. The author has highlighted solutions to the problems. I hope that this chapter will help the reader by providing useful insight on the ESD design and chip development mechanism and will overcome the number of research challenges in order to speed up technological innovation for nanoelectronics in the near future.

In Chapter 2, the synthesis and characterization of poly ethylene-co-vinyl acetate (PEVA) is presented, which plays an important role in solar materials, such as PV modules and agricultural greenhouse polymer sheets. The morphology and miscibility of fabricated nanocomposite films was analyzed and investigated experimentally. It was observed that these nanocomposites showed increased thermal stability, glass transition temperature, high transmittance, and optimum UV-shielding efficiency.

The structural aspects of VOCs using porous silicon coated with a layer of ZnO (PS-ZnO), which is essential in sensing Volatile Organic Compounds (VOCs) is discussed in Chapter 3. Further variations in several parameters were measured and analyzed by experimental means.

Optical tweezers have the potential for application in nanostructure assembly, cancer cell sorting, targeted drug delivery, single-molecule studies, and biosensing. In Chapter 4, a discussion about nano-optical manipulation techniques is presented to solve the problem of difficult integration and diffraction limitations of conventional optical tweezers and holographic optical tweezers, which can be used in biophotonic and biomedical applications.

Chapter 5 is about the most innovative and reliable probe nanolithography techniques based on the spatial confinement of a chemical reaction within a nanometric size frame, which can be applied for controlled deposition of molecules with high precision or nanotransistors that can be used as sensors for biorecognition processes. Interestingly, these motifs can be easily scaled for the macroscale with the use of nanoimprint techniques.

The last Chapter 6, deals with biophotonic technologies, which are used to provide unique and dynamic information about tissue structure and biochemical composition. The authors have highlighted the collaborated clinical research of biophotonic technologies used to characterize brain tissue structure and biochemical composition within the micro to macroscopic regimes in Africa. The advantages and limitations of the present systems have been vastly discussed. Further the authors have emphasized the requirement of smart, simple, and inexpensive solutions to simplify clinical protocols at Point Of Care (POC) activities of the assembly operators, in order to reduce errors and to increase healthcare systems efficiency.

I wish all the authors great success for all their future endeavors and hope that their contribution can help researchers all across the world.

I would like to thank my Author Service Manager Ms. Sara Debeuc for her continuous support, prompt response to all queries, and immense support at every level to complete this project.

Thank you so much Sara.

I acknowledge DST(SR/WOS-A/CS-69/2018) scheme, my mentor Dr. Shrish Tiwari, Bioinformatics, CSIR-CCMB, Hyderabad and finally I would like to thank my husband Mr. Amit Mohan, daughter Arghyaa, son Aryan, and my entire family for their continuous encouragement and understanding during the entire project.

Dr. Ruby Srivastava
Principal Investigator,
CSIR-Center for Cellular and Molecular Biology,
Hyderabad, India

Chapter 1

Electrostatic Discharge, Electrical Overstress, and Latchup in VLSI Microelectronics

Steven H. Voldman

Abstract

Electrostatic discharge (ESD), electrical overstress (EOS), and latchup have been an issue in devices, circuit and systems for VLSI microelectronics for many decades and continue to be an issue till today. In this chapter, the issue of ESD, EOS and latchup will be discussed. This chapter will address some of the fundamental reasons decisions that are made for choice of circuits and layout. Many publications do not explain why certain choices are made, and we will address these in this chapter. Physical models, failure mechanisms and design solutions will be highlighted. The chapter will close with discussion on how to provide both EOS and ESD robust devices, circuits, and systems, design practices and procedures. EOS sources also occur from design characteristics of devices, circuits, and systems.

Keywords: electrical overstress, electrostatic discharge, latchup, system failure, component failure

1. Introduction

Electrostatic discharge (ESD) and electrical overstress (EOS) have been an issue with electrical systems even prior to semiconductors and VLSI technology [1–25]. With the scaling of semiconductor components, ESD continues to be a manufacturing and design issue [2], while latchup also became a reliability and design issue with the introduction of CMOS technology [1–25].

In this chapter, we will address why certain choices and circuit directions are made. This chapter will hopefully provide some insight from a person who has done extensive work in the ESD design and chip development.

In electronic design, a plethora of electrical events can occur. **Figure 1** illustrates the type of topics including ESD, EOS, latchup as well as electromagnetic interference (EMI), and electromagnetic compatibility (EMC). In this chapter, the focus will only be on ESD, EOS, and latchup.

2. Qualification of semiconductor components—electrostatic discharge (ESD)

ESD is a common form of component level failure and system level from manufacturing, shipping, and handling. Historically, the ESD sources in manufacturing

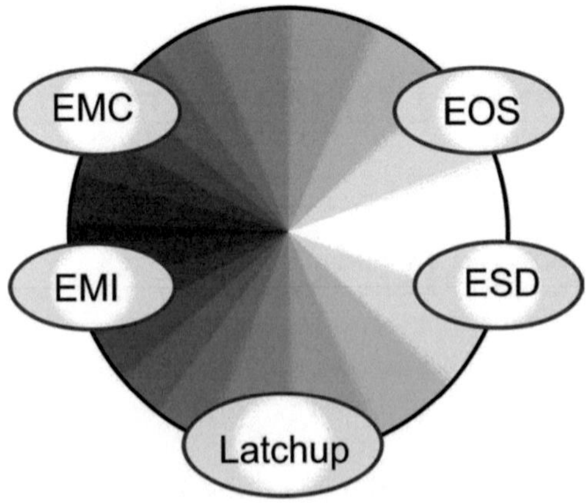

Figure 1.
ESD, EOS, latchup, EMI and EMC.

exceeded the ESD robustness of products leading to component failures. Significant improvements in manufacturing environments as well as ESD solutions in circuit design reduced the concern of component failure levels for many years. Manufacturing charging occurred due to inadequate ionization, material properties, and charging processes from cutting, etching, and dicing. Today, the ESD models performed for qualification and shipping of semiconductor components are as follows [4]:

- human body model (HBM); and
- charged device model (CDM).

With the introduction of VLSI microelectronics, additional qualification of components includes [4]:

- latchup; and
- transient latchup.

CMOS latchup was a large concern in space applications. For low power, CMOS technology was introduced into high density VLSI products. The chapter will address the successes and mistakes made in the industry that lead to latchup failures.

2.1 The human body model (HBM)

Today, the human body model (HBM) is the most established standard for the reliability of components in the semiconductor industry [2–4, 6]. The HBM test is integrated into the qualification and release process of the quality and reliability teams for components in corporations and foundries [6] and is fundamental in the semiconductor chip development cycle.

The human body model (HBM) is regarded as an electrostatic discharge (ESD) event, not an electrical overstress (EOS) event [1–11]. The human body model (HBM) represents a model of a human beings electrical discharge to a semiconductor component. The model assumes that the human being is charged and discharges to the component.

Additionally, HBM failures occur in the power rails and the ESD "power clamps" between the power rails. HBM failures can occur in digital, analog, or radio frequency applications. The failure signature is typically isolated to a single device, or a few elements. HBM failures occur for both positive and negative polarity discharges.

ESD circuits respond to a specific pulse width to mitigate the ESD pulse event. ESD circuits guide the ESD current away from the signal path and to an "alternate current path" through the power grid and ESD power clamp. ESD circuits are "tuned" to be responsive to specific pulse widths.

HBM ESD failures can occur in the following:

- MOSFET active devices;
- diode active devices;
- bipolar junction transistors (BJT) devices;
- resistor passive elements;
- capacitor passive elements;
- inductor passive elements;
- interconnect wiring;
- Vias;
- contacts;
- power rails; and
- decoupling capacitors.

The key question is what is the best choice for providing ESD protection that

- achieves good ESD robustness;
- does not impact circuit functionality;
- does not a reliability concern;
- wide application space for many different designs; and
- migratable to the next technology without significant changes or re-work.

ESD events can be of positive or negative polarity. ESD events can be singular polarity or oscillatory. As a result, ESD protection networks must be able to address both polarity types. In an oscillatory pulse, the positive current flows through the alternate current path through the VDD power grid, and for negative polarity, current flows through the VSS power grid. Additionally, they remain off during normal functional operation of a semiconductor chip.

An example of an ESD protection network is a dual-diode network [3]. The dual-diode ESD network is a commonly used network for complimentary metal oxide semiconductor (CMOS) technology because it has the following essential features:

- low turn-on voltage;
- low capacitance;
- responds to both positive and negative polarity events;
- remains off during functional operation.;
- migratable technology to technology; and
- small area required.

For ESD engineers and technologists, a large advantage of the dual-diode ESD network is that it is easy to migrate from one technology generation to another technology generation. For corporations that used MOSFETs for ESD protection strategies, they generate a large amount of work to be successful, and many times, they are unsuccessful. With MOSFET ESD networks, there is not a scalable strategy that guarantees migration from one generation to another.

Additionally, in CMOS technology, the shallow trench isolation (STI) structure is scalable. In CMOS technology with STI isolation, the breakdown voltage does not change with STI scaling but remains a constant. This allows for suitable migration without issues and without failure or re-work.

The second key advantage is that it has a low turn-on voltage of 0.7 V. The low turn-on voltage allows for discharge of the current through the alternate current loop and redirects the current away from the signal path. The third advantage is that it can be designed with low capacitance, making it suitable for CMOS, advanced CMOS, and RF technologies; this minimizes the impact to chip performance. The fourth advantage is that it does not contain MOSFET gate dielectric failure mechanisms. Unlike MOSFET ESD protection solutions, there is no dielectric, and hence no dielectric failure mechanisms.

A commonly used ESD networks is the grounded gate n-channel MOSFET device [3]. As the technology scales, the MOSFET snapback voltage reduces, leading to an earlier turn-on of the MOSFET. Unfortunately, the dielectric breakdown and overstress are problems in this circuit. Hence, although used in many applications, it is limited in migration to advanced technologies.

A commonly used ESD power clamp is the RC-triggered ESD MOSFET power clamp. Why is it widely used?

- Early turn-on based on the frequency dependent transient impulse from an ESD event.
- Establishes alternate current loop away from the signal path.
- Works well in conjunction with ESD double-diode network on the input node.
- Circuit impedance scales with the MOSFET width in the RC-triggered clamp.

The reason is that the RC-triggered ESD MOSFET power clamp turns on the MOSFET in response to an HBM ESD event. This RC-triggered clamp responds to the ESD current in the power rail leading to turn-on of the MOSFET as opposed to MOSFET breakdown. This element works well with the ESD double-diode network that is used on the signal pins. Additionally, this circuit scales with MOSFET width, and is not dependent on MOSFET second breakdown response.

2.2 Charged device model (CDM)

The charged device model (CDM) is an electrostatic discharge (ESD) test method that is part of the qualification of semiconductor components. The charged device model (CDM) event is associated with the charging of the semiconductor component substrate and package. The charging of the package occurs through direct contact charging, or field-induced charging process (e.g., the field induced charge device model (FICDM)).

The charged device model (CDM) pulse is regarded as the fastest event of all the ESD events [4, 12–15]. Note that the CDM pulse waveform is influenced by the test platform and measurement metrology.

The CDM event has a significantly different characteristic from the HBM event, and requires different ESD circuit solutions. First, the CDM event is oscillatory. Second, the discharge is fast.

CDM event damage occurs in the semiconductor chip through the substrate. It can also occur through the power supply. Charge is stored on the package and the substrate; then, the power supply rapidly discharges through the grounded pin. Since the charge storage is through the entire substrate, it is distributed phenomena and hence spatially dependent. The CDM failure mechanism can be small "pin-hole" in a MOSFET gate structure; this can occur in receiver networks, as well as metal interconnects.

The current path for charged device model (CDM) in components is significantly different from other electrostatic discharge (ESD) events. In the case of the charged device model (CDM), the package and/or chip substrate is charged through a power or ground rail. The component itself is charged slowly to a desired voltage state. As a result, the current flows from the component itself to the grounded pin during ESD testing. This is significantly from other ESD tests (e.g., HBM events) that ground a reference and then apply an ESD event to a signal or power pin. As a result, the current path that a CDM event follows is from inside the component to the pin that is grounded during test.

3. Qualification of semiconductor components—electrical overstress (EOS)

A second area of interest is electrical overstress (EOS). Electrical overstress (EOS) has been an issue in devices, circuits, and systems for VLSI microelectronics for many decades, as early as the 1970s, and continues to be an issue today [1]. Due to a wide variety of pulse events, slow progress has been made at improving EOS robustness.

Electrical overstress (EOS) sources exist from natural phenomena and power distribution [1, 14–25]. Switches, cables, and other power electronics can be a source of electrical overstress. EOS sources exist in devices, circuits, and systems.

3.1 Safe operating area

Figure 2 illustrates the safe operating area (SOA) of a semiconductor device. There is a current limit and a voltage limit on the borders of the SOA. At the corner of the

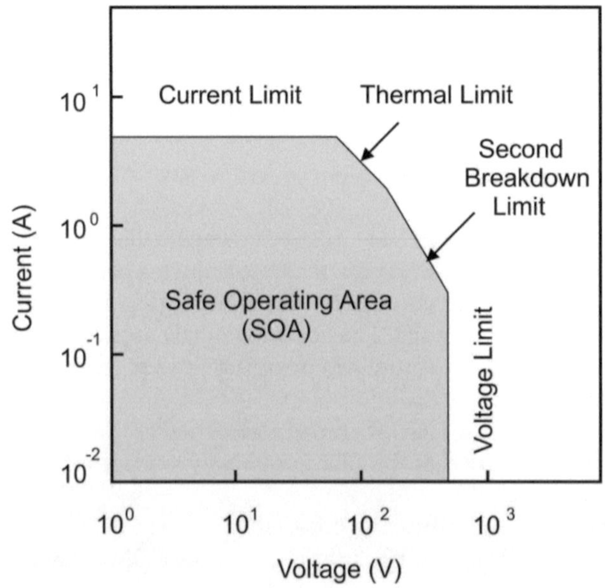

Figure 2.
Safe operating area (SOA).

SOA, the limitation is a thermal limit and a second breakdown limit. The thermal limit has to do with the thermal capacity of a semiconductor device prior to failure. Second breakdown has to do with thermal breakdown limit of a device prior to failure.

Electrical over-voltage (EOV), electric over-current (EOC), and electrical over-power (EOP) can lead to failure mechanisms; these can lead to melted packages, blown single component capacitors and resistors, ruptured packages, blown bond wires, cracked dielectrics, fused and melted metal layers, and molten silicon.

3.2 EOS failure mechanisms

Visual external or internal inspection can be applied to evaluate EOS failure mechanisms. Visual damage signatures can include package lead damage, foreign material, cracks, discoloration, and corrosion. Visual damage can also be evaluated from internal inspection. For internal inspection, visual damage signatures are melted metallurgy, cracked inter-level dielectrics, and molten silicon.

There are certain categories of failures that electrostatic discharge (ESD) does not typically cause and EOS events do cause. Failures that are typically caused by EOS phenomena but not ESD include printed circuit board (PCB), package pin, and wire bond damage.

3.3 EOS protection devices

Electrical overstress (EOS) protection devices are supported by a large variety of technology types. Although the material and device operations may differ between the EOS protection devices, the electrical characteristics can be classified into a few fundamental groups [16–25].

EOS protection networks can be identified depending on whether it is suppressing voltage, as a voltage suppression device, or the current, as a current-limiting device. The voltage suppression device limits the voltage magnitude observed on the signal pins, or power rails of a component, preventing electrical over-voltage (EOV). The current-limiting device prevents a high current from reaching sensitive nodes, hence avoiding electrical over-current (EOC) [1].

Voltage suppression devices can also be sub-divided into two major classifications [1]. Voltage suppression devices can be segmented into devices that remain with a positive differential resistance region and those that undergo a negative resistance region. For a positive differential resistance, these devices can be referred to as "voltage clamp" devices where dI/dV remains positive for all states. For the second group, there exists a region where dI/dV is negative. The first group can be classified as "voltage clamp devices," whereas the second group can be referred to as an "S-type I-V characteristic device," or as a "snapback device." In the classification of voltage suppression devices, the second classification can be associated with the directionality; a voltage suppression device can be "uni-directional" or "bi-directional."

The choice of electrical overstress (EOS) device to use in an application is dependent on the parameters, such as electrical characteristics, cost, and size. The electrical characteristics that are of interest are the breakdown voltage and the forward conduction [1].

The types of voltage suppression devices used electrical overstress (EOS) include transient voltage suppression (TVS) diodes [22], thyristor devices, varistor devices [21], polymer voltage suppression (PVS) devices, and gas discharge tube (GDT) devices [23]. Current-limiting devices can be used in a series configuration for electrical overstress (EOS) protection [16–24]. The choice of the current-limiting EOS protection device is a function of the cost, size, rated current, time response, I^2t value, rated voltage, voltage drops, and application requirements.

4. CMOS latchup

Latchup is a condition where a semiconductor device undergoes a high current state as a result of interaction of a pnp and a npn bipolar transistor [5]. The pnp and npn transistors can be natural to the technology, or parasitic devices. In CMOS technology, these are typically parasitic devices.

When interaction occurs between a pnp and a npn bipolar transistor, or parasitics of CMOS transistors, regenerative feedback between the two transistors can lead to electrical instability. When the two transistors, parasitic or non-parasitic are coupled, the combined device acts as a four-region device of alternating p- and n-doped regions with three physical p-n metallurgical junctions, forming a pnpn structure. When these parasitic pnpn elements undergo a high current state, latchup can initiate thermal runaway and can be destructive.

Latchup events can lead to destruction of a semiconductor chip, package, or system.

There are many reasons why latchup is an issue in today's semiconductor chips. The reason why it is a concern in some corporations differs based on the choices made in the semiconductor technology, latchup design strategy, as well as the latchup methodology.

These are just some of the reasons why today CMOS latchup is not "cured" and remains an issue in today's semiconductor chips. In this text, we will address many of these issues in the future chapters.

5. Conclusions

In conclusion, significant advancements have been made in the understanding of ESD, EOS, and latchup failure mechanisms, as well as solutions to address them have been applied in semiconductor electronics.

Author details

Steven H. Voldman
IEEE Fellow, United States of America

*Address all correspondence to: voldman@ieee.org

References

[1] Voldman S. Electrical Overstress (EOS): Devices, Circuits, and Systems. Chichester: Wiley; 2013

[2] Voldman SESD. Physics and Devices. Chichester: Wiley; 2004

[3] Voldman S. ESD: Circuits and Devices. 2nd ed. Chichester: Wiley; 2015

[4] Voldman S. ESD Testing: From Components to Systems. Chichester: Wiley; 2016

[5] Voldman S. Latchup. Chichester: Wiley; 2007

[6] ANSI/ESD ESD-STM 5.1. ESD Association Standard Test Method for the Protection of Electrostatic Discharge Sensitive Items - Electrostatic Discharge Sensitivity Testing - Human Body Model (HBM) Testing - Component Level. Standard Test Method (STM) Document; 2007

[7] Guest PG, Sikora VW, Lewis B. Bureau of Mines, Report of Investigation 4833. Washington, D.C: U.S. Department of Interior; 1952

[8] Bulgin D. Static electrification. British Journal of Applied Physics, Suppl 2. 1953. DOI: 19.1088/0508-3443/4/s2/333

[9] Martin C. Duration of the Resistive Phase and Inductance of Spark Channels. Atomic Weapons Research Establishment, SSWA/JCM/1065/25. 1996;7:287-293

[10] Hyatt H, Calvin H, Mellberg H. A closer look at the human ESD event. In: Proceedings of the Electrical Overstress/Electrostatic Discharge (EOS/ESD) Symposium. Rome, New York: ESD Association; 1981. pp. 1-8

[11] Hyatt H, Calvin H, and Mellberg H. Bringing ESD into the 20th century. In: Proceedings of the IEEE International Electromagnetic Compatibility (EMC) Symposium; 1982

[12] ANSI/ESD ESD-STM 5.3.1. ESD Association Standard Test Method for the Protection of Electrostatic Discharge Sensitive Items - Electrostatic Discharge Sensitivity Testing – Charged Device Model (CDM) Testing - Component Level. Standard Test Method (STM) Document. Rome, New York: ESD Association; 1999

[13] JEDEC. JESD22-C101-A. A Field-Induced Charged Device Model Test Method for Electrostatic Discharge-Withstand Thresholds of Microelectronic Components; 2000

[14] Renninger R, Jon M, Lin D, Diep T, Welsher T. A field-induced charged device model simulator. In: Proceedings of the Electrical Overstress/Electrostatic Discharge (EOS/ESD) Symposium. Rome, New York: ESD Association; 1989. pp. 59-71

[15] Renninger RG. Mechanisms of charged device electrostatic discharges. In: Proceedings of the Electrical Overstress/Electrostatic Discharge (EOS/ESD) Symposium. Rome, New York: ESD Association; 1991. pp. 127-143

[16] Edison TA. Fuse Block. Washington, D.C: USPTO; US Patent No. 438,305. October 14, 1890

[17] Page CG. Improvement in Induction-Coil Apparatus and in Circuit Breakers. US Patent No. 76,654; 1868

[18] Wright A, Newbury PG. Electric Fuses. 3rd ed. Piscatewey, New Jersey: Institute of Electrical Engineers; 2004. pp. 2-10

[19] Kraz V. Origins of EOS in manufacturing environment. In:

Proceedings of the Electrical Overstress/Electrostatic Discharge (EOS/ESD) Symposium. Rome, New York: ESD Association; 2009. pp. 44-48

[20] Kaschani KT, Gaertner R. The impact of electrical overstress on the design, handling and application of integrated circuits. In: Proceedings of the Electrical Overstress/Electrostatic Discharge (EOS/ESD) Symposium; 2011. pp. 220-229

[21] Philipp HR, Levinson LM. Transient protection with ZnO varistors: Technical considerations. In: Proceedings of the Electrical Overstress/Electrostatic Discharge (EOS/ESD) Symposium. Rome, New York: ESD Association; 1980. pp. 26-34

[22] Hopkins DC. Protective level comparisons for voltage transient suppressors. In: Proceedings of the Electrical Overstress/Electrostatic Discharge (EOS/ESD) Symposium. Rome, New York: ESD Association; 1980. pp. 35-43

[23] Bazarian A. Gas tube surge arresters for control of transient voltages. In: Proceedings of the Electrical Overstress/Electrostatic Discharge (EOS/ESD) Symposium. Rome, New York: ESD Association; 1980. pp. 44-53

[24] Horgan EL. Analytical assessment of electrical overstress effects on electronic systems. In: Proceedings of the Electrical Overstress/Electrostatic Discharge (EOS/ESD) Symposium. Rome, New York: ESD Association; 1980. pp. 140-148

[25] Durgin DL. An overview of the sources and effects of electrical overstress. In: Proceedings of the Electrical Overstress/Electrostatic Discharge (EOS/ESD) Symposium. Rome, New York: ESD Association; 1980. pp. 154-160

Chapter 2

Development and Characterization of Poly Ethylene-Co-Vinyl Acetate (PEVA) Hybrid Nanocomposite Encapsulates for Solar PV

Rashmi Aradhya, Madhu Bilugali Mahadevaswamy and Poornima

Abstract

In the solar photo voltaic (PV) module, encapsulant material provides the environmental protection, insulation, optical absorption, besides serving as a good adhesive between solar cell and components of PV module for improving the efficiency. It is desired to develop an improved encapsulating material by incorporating the light absorbing inorganic nanofillers in thermoplastic polymers. One such matrix material is poly ethylene-co-vinyl acetate (PEVA), finding its importance in solar materials, such as PV modules and agricultural greenhouse polymer sheets. Inorganic nanofillers have the potential to transmit necessary radiance in the UV spectra, which can improve the PV panel efficiency. In this study, the optimum effect of inorganic fillers such as organically modified montmorillonite clay (OMMT) and titanium dioxide (TiO_2) anatase in PEVA matrix is observed. The fabricated nanocomposite films were etched from the glass mold. The morphology and miscibility of fabricated nanocomposite films were analyzed and investigated by scanning electron microscopy (SEM), X-ray diffraction technique (XRD), UV-Vis absorption (UV-Vis), and Fourier-Transform Infrared Spectroscopy (FTIR). The dielectric properties of the fabricated hybrid nanocomposite films were analyzed for its insulation behavior. The thermal behavior was studied using Thermo-gravimetric Analysis (TGA) and Differential Scanning Calorimetry (DSC). The hybrid nanocomposite with 5.0 weight percentage (wt.%) OMMT and 5.0 weight percentage (wt.%) of TiO_2 indicates lowest dielectric constant of 2.4 and marginal increase in dissipation factor with respect to frequency. Increased thermal stability, glass transition temperature, high transmittance and optimum UV-shielding efficiency were found with the same wt.% in the proposed work.

Keywords: characterization, hybrid nanocomposites, poly ethylene-co-vinyl acetate, solar encapsulant, thermo-gravimetric analysis, X-ray diffraction technique

1. Introduction

The necessity to increase the energy generation is to meet the present day scenario of increasing energy demand, environmental pollution and decline in the nonrenewable energy sources. This has led to the explore of various renewable energy resources [1]. Even though renewable energy sources has appeared to be the best possible solution in terms of sustainability, portability, and availability, it is not commercially feasible in short period of time as compared to conventional fuels. This is due to complex processing technologies and material supply limitations. The solar

The photovoltaic (PV) provides pure and renewable alternative energy source, potentially advantageous to the environment by preventing greenhouse gases being generating and go into the atmosphere. The PV industry is increasing quickly as the demand for cleaner energy worldwide continues to increase.

As the industry develops, it is significant that suitable materials are available to meet the several requirements such as robustness, performance, cost and global availability.

Silicone materials have been used in a wide variety of applications in industries, such as construction and electronics industries, and perfect product to meet the requirements in the PV module assembly market.

In order to achieve best possible performance, the encapsulant material used in the development of photovoltaic modules should satisfy a number of requirements, which include: high optical transmittance of incident light, good dielectric properties (thermally conductive and electrically insulating), good mechanical strength to protect the PV cells from outside mechanical loads and thermal stresses, good adhesion to both glass and PV cells, and sufficiently robust to survive 20–30 years in the field. This paper provides the overview and the key requirements for materials as PV encapsulants. In the recent developments of hybrid materials, both organic-inorganic nanofillers are receiving major attention due to broad range of potential applications [2].

PEVA, as used in the solar industry, is a thermoplastic elastomer that is formulated with a curing agent, UV absorbers, as well as photo and thermo-antioxidants. Even though, PEVA encapsulation meets largely the rigorous material property necessity at an attractive price, there exists a couple of areas for improvement.

Composites of PEVA reinforced with OMMT and TiO_2 nanofillers have emerged as a key field of research because of numerous advantages, which includes an improvement in mechanical, dimensional, thermal properties and enhanced transparency when dispersed nanoclay platelets suppress polymer crystallization [3]. Even at very low weight percentage of the nanofiller addition, capable to alter properties of the polymer present in nanocomposite compared to the pristine polymer because the bulk of polymer chains positioned such that they are in close contact with the OMMT surface. OMMT has outer tetrahedral layers that contain Si^+ and O^- atoms. These outer layers align themselves due to Van der Waals forces of attraction between them to form nonionic bond [4].

Nanocomposites developed using organic polymer and inorganic nanofillers forms a new set of materials with confirmation of improved performance when compared with their macroscopic counterparts [5]. The nanocomposite film developed combines the unique properties of inorganic components with the processability of polymer in one material and makes them attractive in varieties of useful applications [6]. Polymers are considered to be most appropriate accommodating matrices for composite materials because they can be easily tailored by adding inorganic filler which has long-term stability and good processability to suit the variety of physical properties. Inorganic nanofillers exhibit better optical, catalytic, electronic and magnetic properties, which are considerably different from their bulk states. The addition of inorganic nanofillers to the polymer matrix allows unique physical properties as well as the implementation of new end-user features. Depending on the particle size, particle shape, specific surface area and chemical nature of the nanofillers, intended polymer matrix properties can be modified [7]. A good cross-linking agent is required for better interactions between the polymer matrix and the nanofillers, as it is directly linked to morphology, crystallinity, photothermal stability, shelf life and thermal stability and also it affects the curing rate by generating the free radical through thermal decomposition [8, 9].

In the present work, pristine PEVA and nanocomposite films with OMMT and TiO_2 as nanofillers with dicumyl peroxide as a curing agent were fabricated by solution casting technique which is found to the optimized fabrication method. From the exhaustive literature review, it was also concluded that dicumyl peroxide is the optimized curing agent for EVA polymer and synthesis of nanocomposite films [10–12].

The nanocomposite films were characterized by Fourier Transform Infrared spectroscopy (FTIR), X-ray diffraction (XRD), Thermogravimetric analysis (TGA), UV-Vis spectroscopy, scanning electron microscopy (SEM) and electrical parameters with LCR meter. The interaction between EVA and OMMT and TiO_2 nanofillers was investigated and the results demonstrate the doping effect of OMMT and TiO_2 nanofillers. The objective of the study carried out to determine the effects of OMMT and TiO_2 nanofillers on PEVA polymer and to investigate introduction of inorganic TiO_2 nanofiller will influence dispersion of the nanoparticles in the polymer matrix for encapsulant application in solar PV modules with improved thermal stability, dielectric properties and optimum UV-shielding efficiency. In the present work, the investigations were carried out on the electrical, thermal and morphological properties of OMMT and TiO_2 nanofillers added in PEVA.

2. Experimental

2.1 Materials

The base matrix material used is poly (ethylene-co-vinyl acetate) vinyl acetate 25 wt.%, melt index 19 g/10 min (190°C/2.16 kg), contains 200–900 ppm BHT as an inhibitor. Nanofillers such as nanoclay (montmorillonite clay) surface modified: 25–30 wt.% trimethyl stearyl ammonium and titanium (IV) oxide-anatase (nanopowder, <25 nm particle size, 99.7% trace metals basis). Dicumyl peroxide 98% bis (α-dimethyl benzyl) peroxide is added to the composite as a curing agent. All materials were supplied by Sigma-Aldrich Pvt. Ltd., Bangalore. Toluene with 98% purity used to dissolve the polymer and nanofillers were procured from Karnataka Fine Chem., Bangalore.

2.2 Composites sample preparation

The good dispersion is a key challenge to achieve the best possible combination of matrix nanoparticles. For that reason, a new dispersion technique, such as ultra-sonication, was followed by manual mixing with different mixing speed and time. This processing method was optimized because it was not very complicated from laboratory processing point of view and commercially available polymers and nanoparticles could be mixed easily and ease to prepare a composite sample. The agglomeration of nanosized particles was significantly reduced with this mixing technique resulting in a well dispersed and homogeneous mixture of nanocomposites.

Pristine PEVA with curing agent dicumyl peroxide (DCP) doped with OMMT and TiO_2 nanofillers, at different concentrations were prepared by the solution casting method as shown in **Figure 1**. Initially, the PEVA solution was prepared by weighing the appropriate amount of PEVA and dissolving it in toluene at 45°C with the help of magnetic stirrer for 45 min and for the solution 2, 5, 7 wt.%,

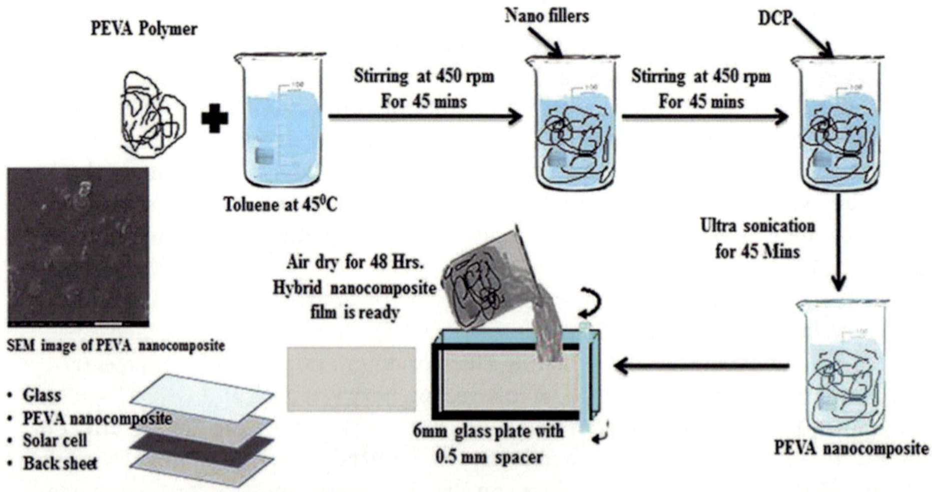

Figure 1.
Process followed for making composite films.

of OMMT and TiO_2 nanofillers, were mixed and stirred vigorously by using a magnetic stirrer until it is uniformly dispersed, a clear multi-component PEVA dispersion was obtained. Further, curing agent was added to the solution and ultra-sonication was carried out for 45 min to achieve the complete dispersion of nanofillers in PEVA solution. Appropriate mixtures of PEVA and nanofillers solution were poured on to the clean glass plate and left to dry at room temperature for about 24 h. The fabricated nanocomposites were post cured for 48 h in a hot air oven at 45°C then the nanocomposite films were removed from the glass plate and cut into required sizes for characterization. The thickness of the obtained composite films was measured using a digital micrometer at different places and average value was taken.

2.3 Characterization

Electrical parameters measurement were carried out for pure PEVA and PEVA nanocomposite samples by using fully automated high-precision four-terminal LCR meter (HIOKI-IM3536) and a four-terminal probe (HIOKIL2000) in the frequency range from 10 Hz to 4 MHz at a room temperature. The capacitance (c), dissipation factor (tanδ) and ac resistance were recorded directly and from which the dielectric constant (ε') and dielectric loss (ε'') were calculated.

The surface morphology of the PEVA and PEVA nanocomposite samples has been observed and recorded by using a scanning electron microscopy (SEM (JEOL, JSM-ITLV model)) with an accelerating voltage of 10 kV. Surface fractured samples were gold coated before making the observations.

XRD analysis of the nanocomposite films was carried out using powder X-ray diffractometer (Shimadzu-7000 with Cu-Kα radiation). Fourier Transform Infrared Spectra (FTIR) was recorded from the FTIR (Bruker-alpha) and analyzed for metal oxide bond stretching. The UV absorbance spectrum of the samples was measured using the Agilent Cary 60, Version 2.00, UV-Vis spectrophotometer with the scan rate of 60 (nm/min) with dual beam mode.

The thermal stability of nanocomposite films was analyzed using a thermogravimetric analyzer, TA Instrument, Q600, with a heating rate of 10°C/min with a temperature ranging from 30 to 800°C in the nitrogen gas atmosphere. The formulation and identification of the samples are listed in **Table 1**.

Sample identification and formulation	PEVA + DCP (wt.%)	OMMT (wt.%)	TiO_2 (wt.%)
PEVA + DCP	100	—	—
PEVA + DCP + 5 wt.% OMMT	95	5	—
PEVA + DCP + 5 wt.% OMMT + 5 wt.% TiO_2	90	5	5

Table 1.
Sample identification and formulation.

3. Results and discussion

3.1 XRD analysis

X-ray diffraction spectrum analysis is useful technique to evaluate the intercalation and exfoliation layered structure arrangement of OMMT and TiO_2 nanofillers in the PEVA base polymer matrix. **Figure 2** shows the complete XRD spectrum obtained for the nanocomposite samples from 5 to 50°. The shift in the characteristics peak of PEVA indicates the existence of good bonding between the nanofillers and pristine PEVA matrix [13]. Diffraction patterns obtained from fabricated PEVA + 5 wt.% OMMT + 5 wt.% TiO_2 nanocomposite showed broader peaks at lower value (6.44°) and at higher value (25.68°), suggesting intercalation/exfoliation of PEVA molecules into the intergallery spacing of OMMT. Peaks with similar diffraction patterns reported in the literature [3, 14] confirmed the good bonding between fillers and the polymer. XRD pattern also indicates the semi-crystalline/amorphous nature of the nanocomposite films. From the characteristics peaks, using Bragg's law, d-spacing can be calculated. From the result, it was observed that prominent diffraction peak obtained with OMMT nanofiller at $2\theta = 6.47°$, correspondingly intergallery d-spacing varies from 18.5 to 24.7 Å [15, 16] and for TiO_2 anatase [17, 18] diffraction peak occurred in the wide angle range of 2θ ascertained that the peaks at 25.36° shows crystalline structures of anatase synthesized TiO_2 nanoparticles. The complete or higher degree of exfoliation of layered crystalline structure of nanofillers in polymer matrix indicates the absence of corresponding peaks in the XRD spectra of the nanocomposite samples. The absence of peaks obtained from the XRD spectrum may be due to several reasons: Firstly, a very small or low concentration of the nanofillers in the regions, where X-ray beams scans the materials and secondly, loss of symmetry in certain crystallographic direction.

3.2 Scanning electron microscopy

Figure 3(a)–**(c)** shows the SEM images of PEVA, PEVA + DCP + 5 wt.% OMMT and PEVA + DCP + 5 wt.% OMMT + 5 wt.% TiO_2 nanocomposite films.

From the **Figure 3(b)** and **(c)**, it was found that the nanofillers were uniformly distributed and good adhesion between PEVA polymer and nanofillers. From the observation of surfaces in SEM micrographs **Figure 3(c)**, the possible agglomeration of TiO_2 nanofiller was ruled out and confirms the well-dispersed nanofillers in the PEVA polymer matrix. The characteristic peaks corresponding to OMMT and TiO_2 present in the XRD pattern of the nanocomposite films, referred to **Figure 2** supports a well-structured nanocomposite formation.

3.3 Fourier Transform Infrared Spectra (FTIR) analysis

Figure 4 shows the FTIR spectrum obtained by transmittance technique to find the different functional groups present and the interactions among nanofillers,

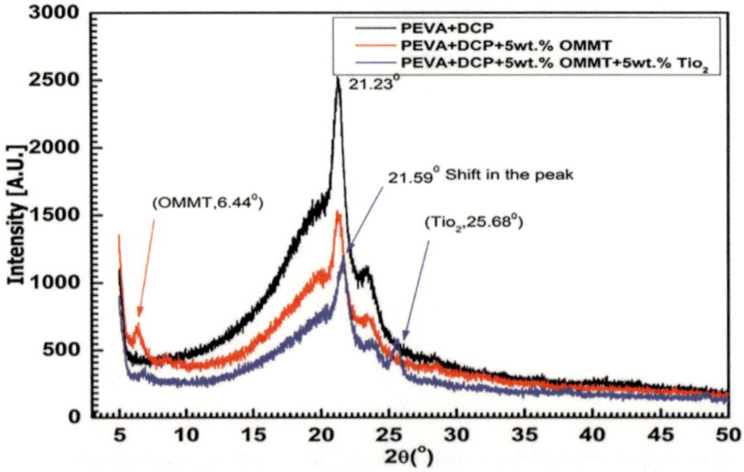

Figure 2.
XRD spectra of PEVA, PEVA + OMMT + TiO_2 nanocomposites.

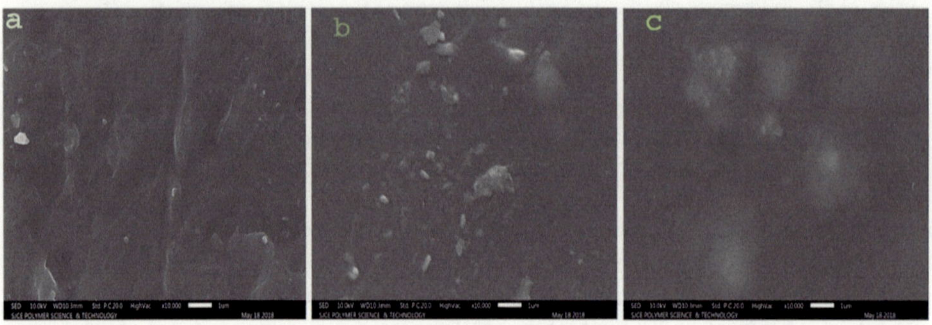

Figure 3.
SEM image of: (a) PEVA + DCP, (b) PEVA + DCP + 5 wt.% OMMT, and (c) PEVA + DCP + OMMT + TiO_2 nanocomposites.

curing agent and PEVA matrix. The effects of OMMT + TiO_2 nanofillers on the molecular structure and crystallization behavior of the PEVA composites can also be investigated. The FTIR spectrum of PEVA + OMMT + TiO_2 composite films shows the presence of the characteristic peaks of pure PEVA and pure OMMT and TiO_2. However, there is a relative shifting of the transmittance bands with increasing nanofillers content, due to hydrogen bonding interactions between -CH_3 groups of PEVA.

The FTIR spectra of nanocomposite films show the characteristic transmittance bands of pristine PEVA and OMMT + TiO_2 nanofillers. It was observed that the pre-dispersing procedure carried out in the preparation of the nanocomposite can considerably alter the morphology of the pure PEVA. The transmittance value of nanocomposite films found to be increased from 400 to 900 cm^{-1} as compared with the neat PEVA sample. The peaks at 720.55, 609.38, 514.67, and 463.15 cm^{-1} belongs to the characteristic peaks of OMMT + TiO_2 nanofillers, which are caused due to stretching vibrations and flexural vibration of O-Ti-O. The absorption peaks at around 1646 cm^{-1} and the wide peaks at around 2950.14 cm^{-1} are associated to flexural vibration of H-O-H bonds of physical inclusion of water and surface Ti-OH bonds and hydrogen bonded molecular water species respectively. This result indicates that a large number of -OH groups were absorbed on the surface of OMMT + TiO_2 nanofillers, which forms hydrogen bonds with the PEVA. In

PEVA + DCP + 5 wt.% OMMT + 5 wt.% TiO_2 sample has the stretching vibration of hydroxyl (-OH) group at 3456 cm^{-1}, indicating its hydrolyzation into silanol with hydroxyl (-OH) group. The stretching vibration of vinyl C=C appeared at 1600 cm^{-1}, out-of-plane wagging peak of vinyl CH_2 at 962 cm^{-1}, and the characteristic peak of the Si-O bond at 463.15 cm^{-1}. The abundance amount of free -OH groups are available on the nanoparticle surface and the density of hydrogen bonded PEVA segments will be high at the region just close to the nanoparticle surface. This process of formation of hydrogen bonds in the PEVA nanocomposite indicates the formation of two nanolayer; the tightly and strong nanolayer of PEVA segment at interface region with nanoparticles (**Table 2**).

3.4 UV-Vis spectroscopy

UV-Vis spectroscopy is used to understand the optical response of a polymer and nanocomposites when subjected to the absorption of UV-Vis radiation of the electromagnetic spectrum. In commonly used greenhouse covers and solar PV encapsulant, the thermal conductive and electrically insulative fillers are used to improve insulating properties which tends to reduce the transmission of visible

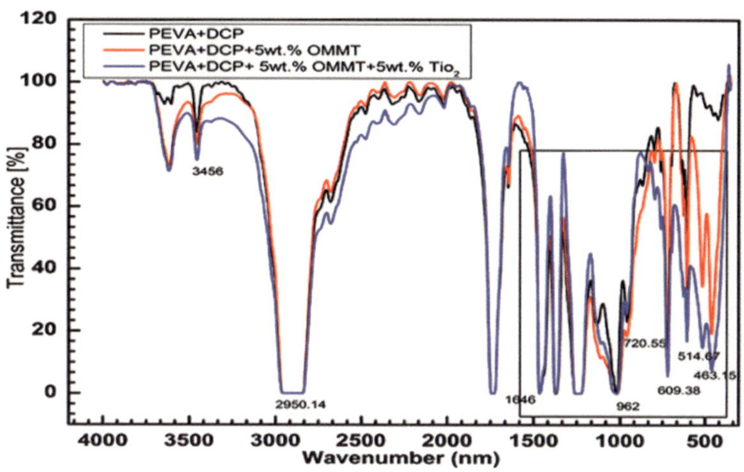

Figure 4.
FTIR spectra of PEVA + OMMT + TiO_2 nanocomposites.

Wave number (cm^{-1})	Functional groups
463.15	O-Ti-O
514.67	
609.38	
720.55	
962	Vinyl CH_2
1600	Vinyl C=C
1646	H-O-H
2950.14	H-O-H
3456	-OH

Table 2.
FTIR peaks corresponding to the functional groups.

light through the film. The ultraviolet and visible infrared spectra obtained for nanocomposite films are shown in **Figure 5**. It could be elucidated from the spectrum that it has a higher transparency/lower absorbance window between 200 and 800 nm. Absorption peaks are noticed between 200 and 280 nm for all samples. As the OMMT nanofiller is added, it is observed that the intensity of the absorption is augmented. The vertical axis shows the absorbance indicating the amount of light absorbed by the samples. The higher value in the vertical axis indicates the quantity of particular wavelength being absorbed.

They confirmed a decrease in UV transmission of PEVA and retained a high visible light transmission. Therefore, these nanocomposite films with light selective properties would be an excellent component for PV encapsulation. Among the nanocomposite films, the high absorption is observed below 230 nm which is not taken into consideration because of the large absorption coefficient of C=O bonds due to $\pi \rightarrow \pi^*$ transition (absorbance value exceeded 2, below 230 nm). Nanocomposite film with 5 wt.% OMMT and 5 wt.% TiO_2 allowed the maximum transmission of light and performs better.

3.5 Thermogravimetric analysis (TGA)

Thermo-gravimetric analysis (TGA) is usually used to understand the thermal stability behaviors of polymeric materials. Thermo gravimetric analysis (TGA) is finding increasing applications for investigations on the pyrolysis and combustion behavior of polymers. When a sample is subjected to TGA, decomposition occurs at a very slow rate until a critical temperature is reached. The pyrolysis rate then increases very rapidly to a maximum, leading to complete combustion and then the rate drops rapidly. Such a behavior is characteristic of a large number of decomposition processes, including pyrolysis of many polymers.

The TGA graph of fabricated nanocomposite samples shown in **Figure 6**.

The thermal degradation of EVA shows two distinct phases, which have been assigned to the loss of acetic acid and the degradation of the resulting unsaturated material. It consists of a combination of deacetylated vinyl acetate or unsaturated entities and ethylene entities of PEVA. From the literature, it was found that, as the process of deacetylation continues, inert degradation of polyene, oxidative degradation of

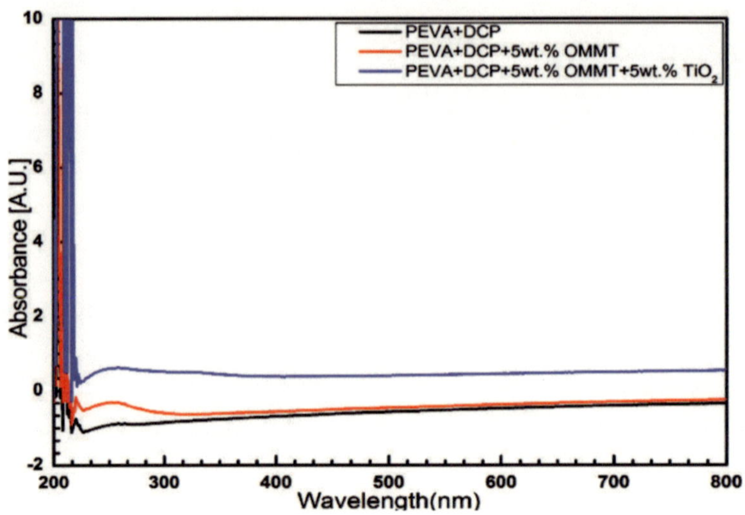

Figure 5.
UV-Vis spectra of PEVA and nanocomposites.

polyene [13, 14, 19, 20] taken place. Deacetylation occurs for all the samples in the temperature region between 300 and 400°C. After an inert or oxidative deacetylation of PEVA, a polyene is formed, further inert degradation of this polyene by chain scission processes, which could be neglected since for PV encapsulant application the maximum operating temperature is around 100°C [20]. The TGA thermographs assured the degradation of the hybrid nanocomposite at a higher temperature (around 420°) with faster rate as high transparency was observed form the UV-Vis spectrum.

3.6 Dielectric constant (ε′)

Figure 7 shows the variation of dielectric constant (ε′) with the frequency and loading of OMMT and TiO_2 nanofillers. The dielectric constant decreases up to 10^4 Hz above which it increases over the measured frequency range up to 10^6 Hz in neat PEVA, PEVA + 5 wt.% OMMT and PEVA + 5 wt.% OMMT + 5 wt.% TiO_2 nanocomposite samples. The decrease of dielectric constant with frequency in neat PEVA may be attributed to the fact that at a low-frequency dielectric constant for the polar material is due to the contribution by various polarizations namely, electronic, ionic, orientational and interfacial polarization. The resulting total polarization in dielectric materials is due to sum of these four types of polarization [21]. The polarization contribution due to dipole orientation dominates at low frequencies.

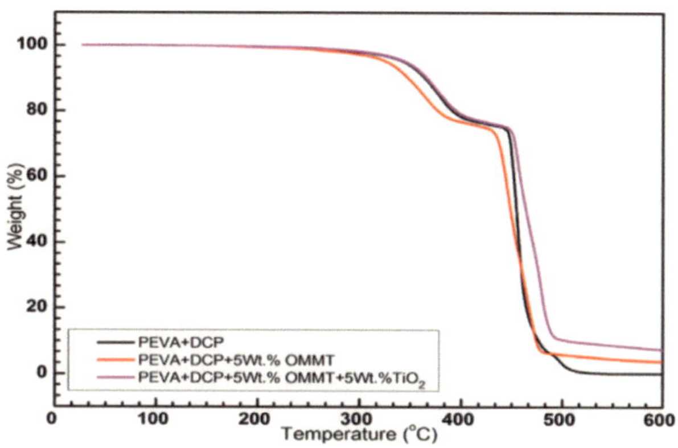

Figure 6.
TGA thermograms of PEVA and nanocomposites.

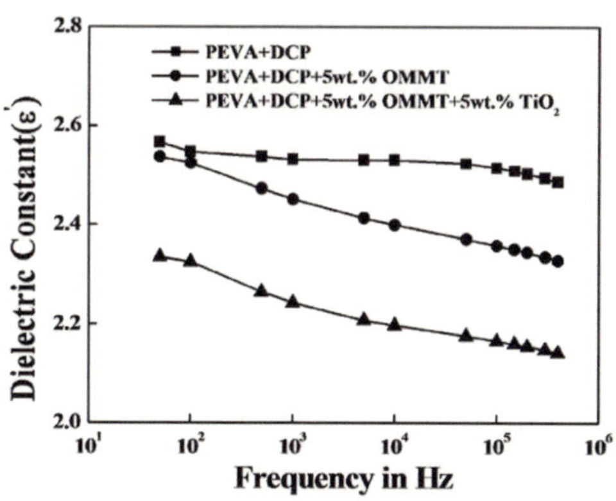

Figure 7.
Plot of dielectric constant of neat PEVA, PEVA + OMMT and PEVA + OMMT + TiO_2 nanocomposites with frequency.

The speed of dipole rotation at high frequency is insufficient to match the shift in the applied AC bias [22].

The plot of dielectric constant (ε′) with frequency of PEVA + 5 wt.% OMMT nanocomposite might be understood by that the clay in exfoliated/intercalated form within the polymer matrix has terminated the regular long chains of the polymer and the resulting polymer structure is more random with shorter chains. Within the polymer due to nanoconfinement of the clay, dipole orientation may be constrained and not easily moved. The nanoscopic confinement effect from layered silicate inorganic hosts was reported by Anastasiadis et al. [23]. The author reported that X-ray diffraction showed that the polymer is confined within 1.5–2.0 nm. This confinement effect is directly reflected in the local reorientational dynamics. We believed that the orientation polarization is thus largely reduced due to this randomly distributed and confined structure within the polymer. It is observed that with the addition of OMMT nanofillers the dielectric constant in PEVA polymer decreases. The changes in the dielectric constant values of polymer clay nanocomposites (PCNs) have a strong correlation with intercalated/exfoliated structures of OMMT in the polymer matrix [24, 25]. It has been established that for the PCNs, the dielectric constant decreases due to the predominance of exfoliated-OMMT structure in a polymer matrix, whereas it increases for a large amount of intercalated structure [25–29].

It is also evident from **Figure 7**, that with the addition of 5 wt.% TiO_2 nanofiller to PEVA-OMMT nanocomposite sample with 5 wt.% OMMT, the dielectric constant increases. This increase in dielectric constant of nanocomposite sample with the addition of 5 wt.% TiO_2 is due to TiO_2 exhibits strong ionic polarization due to Ti^{4+} and O^{2-} ions and hence PEVA nanocomposite with TiO_2 filler has higher value of dielectric constant [30]. The polarization mechanism in nano-TiO_2 and OMMT over the frequency studied are closely identical, and therefore the resulting trends in the dielectric constant variations in both the nanocomposites are expected to be similar. The effective dielectric constant with TiO_2 is always higher than the values obtained with OMMT as nanofiller. This result may be due to (i) nanofiller inherent dielectric constant, and (ii) nanofiller concentration. TiO_2 has a higher dielectric constant than OMMT and hence the effective dielectric constant increases with the addition of TiO_2 and is always higher than PEVA + 5 wt.% OMMT nanocomposites. The other aspect is, since the density of OMMT (density $\approx$ 1.9 g/cm^3) is less than that of TiO_2 (density $\approx$ 4.2 g/cm^3), for the same quantum of loading, the PEVA nanocomposite with OMMT as a nanofiller will have more number of OMMT nanoparticles as compared to TiO_2 nanoparticles. With fillers in nanometer scales, this difference in the number of particles can be very significant. Thus, for a fixed filler loading, OMMT nanofillers in the PEVA nanocomposite films will introduce more interfaces causing additional restrictions to the polymer chain mobility when compared to TiO_2 nanofillers. This enhanced chain mobility restrictions with OMMT nanofillers coupled with the effect of a lower OMMT dielectric constant, will lead to lowering of effective dielectric constant of PEVA + 5 wt.% OMMT nanocomposite, to values lower than that of the TiO_2 filled ones. Although both these processes will be dynamic in the nanocomposites and also it is difficult to segregate their individual contributions. Further, the variations in dielectric constant are not linear with respect to filler loading. However, there are indications of existence of a threshold value of filler loading at which the dielectric constant shows maximum variation with respect to base polymer PEVA values [31].

3.7 Dissipation factor (tanδ)

Figure 8 shows the variation of dissipation factor (tanδ) of neat PEVA, PEVA + 5 wt.% OMMT, and PEVA + 5 wt.% OMMT + 5 wt.% TiO_2 nanocomposites

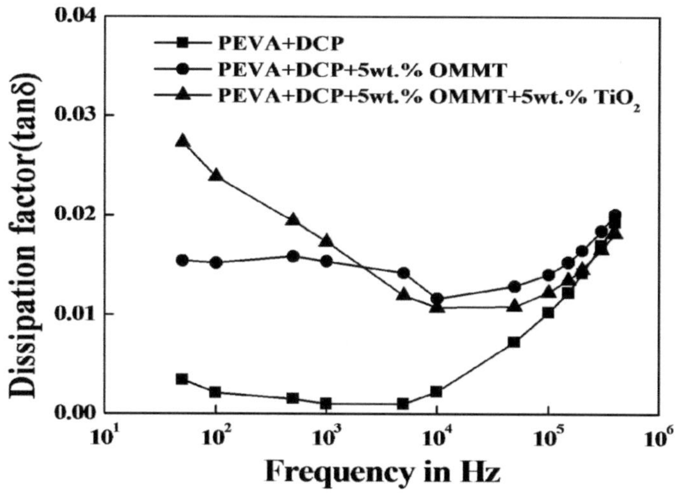

Figure 8.
Plot of dissipation factor (tanδ) of neat PEVA, PEVA + OMMT and PEVA + OMMT + TiO_2 nanocomposites with frequency.

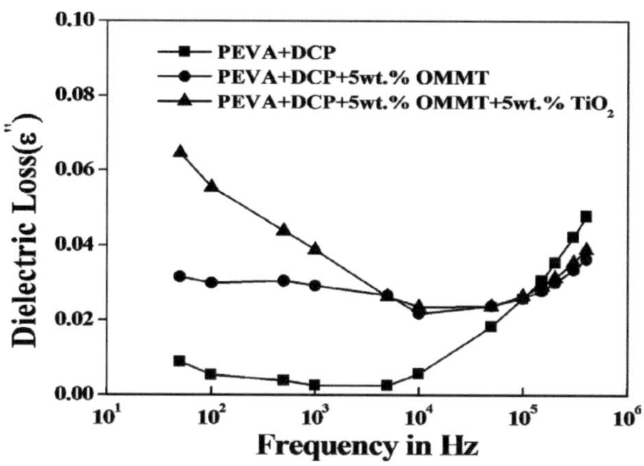

Figure 9.
Plot of dielectric loss (ε") of neat PEVA, PEVA + 5 wt.% OMMT and PEVA + 5 wt.% OMMT + 5 wt.% TiO_2 nanocomposites with frequency.

with respect to frequency. The dissipation factor decreases with increase in frequency in low-frequency range up to 10^4 Hz and above which it increases with increase in frequency. Though there is an increase in tanδ in a high-frequency range beyond 10^4 Hz the dissipation factor values are well within the acceptable range and no substantial increase. It is observed that with the addition of nanofillers the dissipation factor increases in PEVA polymer.

The dissipation factor variations of PEVA nanocomposites are dependent on the frequency of the applied voltage and the temperature of measurement. Therefore the temperature of the measurement was maintained constant. The dissipation factor depends on the electrical conductivity of polymer nanocomposites. The electrical conductivity in turn depends on the number of charge carriers in the bulk of the material, the relaxation time of the charge carriers and the frequency of the applied electric field. Since the measurement temperature is maintained constant, its influence on the relaxation times of the charge carriers is neglected. Over the range of frequency used, charge transport therefore is dominated by the lighter electronic charge carriers.

3.8 Dielectric loss (ε′)

Figure 9 shows the variation of dielectric loss (ε″) of neat PEVA, PEVA + 5 wt.% OMMT and PEVA + 5 wt.% OMMT + 5 wt.% TiO_2 nanocomposites with frequency and nanofillers. It is observed that increase in dielectric loss with the addition of nanofillers in PEVA polymer and the increases with frequency beyond 10^5 Hz in neat PEVA, PEVA + 5 wt.% OMMT and PEVA + 5 wt.% OMMT + 5 wt.% TiO_2 nanocomposites. Beyond 10^5 Hz dielectric loss in nanocomposites has reduced as compared to that of neat PEVA. Loss factor in nanocomposites might be the contribution from dipole orientation, conduction loss and interfacial polarization [32]. The measurement results of dielectric loss are helpful to confirm the polarization mechanism. The loss factor ε″ of composite systems could be expressed as a sum of three distinct contributions [32], that is,

$$\varepsilon'' = \varepsilon''_{dc} + \varepsilon''_{MW} + \varepsilon''_{D} \tag{1}$$

where ε''_{dc} is due to conduction loss contribution, ε''_{MW} is due to interfacial polarization (Maxwell-Wagner) contribution and ε''_{D} is the dipole orientation or Debye loss factor. The conduction loss and interfacial polarization loss factors are expressed as $\varepsilon''_{dc} = \sigma_{dc}$.

$$\varepsilon''_{dc} = \frac{\sigma_{dc}}{2\pi f \varepsilon_0} \tag{2}$$

$$\varepsilon''_{MW} = \varepsilon_{\infty}\left(1 + \frac{K}{1 + (2\pi f)^2 \tau^2}\right) \tag{3}$$

ε_{∞} and K are calculated considering two different dielectric constants of the sample at the interfaces and τ is the relaxation time of the interfacial polarization. ε_o and f are the dielectric constant of vacuum and frequency, respectively. By expressing Eqs. (2) and (3) in logarithmic form and plotting log ε''_{dc} and log ε''_{MW} versus log f, we can obtain two different curves: the log ε''_{dc} against log f represents a straight line and the log ε''_{MW} against log f represents a sigmoidal curve. From **Figure 9**, it is evident that, at high frequencies, a broad loss peak appears. This broad loss peak corresponds to the dipole orientation polarization. At low frequency-range, the plot shows a straight line indicating the existence of conduction loss.

4. Conclusions

This study shows that a fair degree of success has been achieved in production of pristine PEVA base and organically modified nanoclay (OMMT) and TiO_2 added nanocomposite films. The pristine PEVA and organically modified nanoclay (OMMT) and titania (TiO_2) added nanocomposite films have been analyzed by series of tests. The following conclusions were drawn from the research work:

i. The XRD results indicate that the absence of the characteristic peak indicates the exfoliation of the OMMT platelets in the Epoxy matrix.

ii. The SEM micrographs indicate that the nanofillers were uniformly distributed and good adhesion between PEVA polymer and nanofillers.

iii. The FTIR spectra of nanocomposite films show the characteristic transmittance bands of pristine PEVA and OMMT + TiO_2 nanofillers. The existence of a chemical interaction (formation of hydrogen bond) between nanofillers and polymer chains of the PEVA and nanocomposite films has been established through Fourier Transform Infrared spectroscopy.

iv. From the UV-Vis, it was confirmed that there is a decrease in UV transmission of PEVA and retained a high visible light transmission. Therefore, these nanocomposite films with light selective properties would be an excellent component for PV encapsulation.

v. The TGA thermographs assured the degradation of the hybrid nanocomposite at a higher temperature (around 420°) with faster rate as high transparency was observed from the UV-Vis spectrum.

vi. The results of dielectric constant and tanδ of nanocomposites with respect to frequency, nanofiller concentrations, were interesting and intriguing. The frequency and filler dependant dielectric constant characteristics of PEVA and nanocomposites films suggest that there is an inhibition in the mobility of PEVA polymer chains in the bulk of the nanocomposites.

vii. The dissipation factor variation with respect to frequency and filler loading indicate that there is probably a reduction in electrical conductivity in nanocomposites. This is mainly due to inhibition in the mobility of available charge carriers. Though there is an increase in tanδ in a high-frequency range beyond 10^4 Hz the dissipation factor values are well within the acceptable range and no substantial increase.

viii. Some encouraging results such as decrease in dielectric loss, dissipation factor and dielectric loss with the addition of nanofillers, indicates that the nanocomposite films can be used as solar encapsulants. The UV-vis result also indicates that the nanocomposites films are transparent to light in the range between 200 and 800 nm and insights are gained for understanding of the polarization mechanisms involved.

Author details

Rashmi Aradhya*, Madhu Bilugali Mahadevaswamy and Poornima

Department of Electrical and Electronics Engineering, Siddaganga Institute of Technology, Tumkur, Karnataka, India

*Address all correspondence to: rash_mysore@sit.ac.in; rash_mysore@yahoo.com

References

[1] I. E. Agency. World Energy Outlook 20132013. DOI: 10.1787/weo-2013-en

[2] Hanemann T, Szabó DV. Polymer-nanoparticle composites: From synthesis to modern applications. Materials. 2010;**3**(6):3468-3517. DOI: 10.3390/ma3063468

[3] Duquesne S, Jama C, Le Bras M, Delobel R, Recourt P, Gloaguen JM. Elaboration of EVA-nanoclay systems—Characterization, thermal behaviour and fire performance. Composites Science and Technology. 2003;**63**(8):1141-1148. DOI: 10.1016/S0266-3538(03)00035-6

[4] Alexandre M, Dubois P. Polymer-layered silicate nanocomposites: Preparation, properties and uses of a new class of materials. Materials Science & Engineering R: Reports. 2000;**28**(1):1-63. DOI: 10.1016/S0927-796X (00)00012-7

[5] Jeon IY, Baek JB. Nanocomposites derived from polymers and inorganic nanoparticles. Materials. 2010;**3**(6):3654-3674. DOI: 10.3390/ma3063654

[6] Kango S, Kalia S, Celli A, Njuguna J, Habibi Y, Kumar R. Surface modification of inorganic nanoparticles for development of organic-inorganic nanocomposites—A review. Progress in Polymer Science. 2013;**38**(8):1232-1261. DOI: 10.1016/j.progpolymsci.2013.02.003

[7] Lü C, Yang B. High refractive index organic-inorganic nanocomposites: design, synthesis and application. Journal of Materials Chemistry. 2009;**19**(19):2884. DOI: 10.1039/b816254a

[8] Tcherbi-Narteh A, Hosur M, Triggs E, Owuor P, Jelaani S. Viscoelastic and thermal properties of full and partially cured DGEBA epoxy resin composites modified with montmorillonite nanoclay exposed to UV radiation. Polymer Degradation and Stability. 2014;**101**(1):81-91. DOI: 10.1016/j.polymdegradstab.2013.12.033

[9] Agroui K, Collins G, Oreski G, Boehning M, Arab AH, Ouadjaout D. Effect of crosslinking on EVA-based encapsulant properties during photovoltaic module fabrication process. Revue des Energies Renouvelables. 2015;**18**(2):303-314

[10] Fernández AI, Haurie L, Formosa J, Chimenos JM, Antunes M, Velasco JI. Characterization of poly(ethylene-co-vinyl acetate) (EVA) filled with low grade magnesium hydroxide. Polymer Degradation and Stability. 2009;**94**(1):57-60. DOI: 10.1016/j.polymdegradstab.2008.10.008

[11] Mishra SB, Luyt AS. Effect of organic peroxides on the morphological, thermal and tensile properties of EVA-organoclay nanocomposites. Express Polymer Letters. 2008;**2**(4):256-264. DOI: 10.3144/expresspolymlett.2008.31

[12] Zhang W, Chen D, Zhao Q, Fang Y. Effects of different kinds of clay and different vinyl acetate content on the morphology and properties of EVA/clay nanocomposites. Polymer. 2003;**44**(26):7953-7961. DOI: 10.1016/j.polymer.2003.10.046

[13] Wang KH, Choi MH, Koo CM, Choi YS, Chung IJ. Synthesis and characterization of maleated polyethylene/clay nanocomposites. Polymer. 2001;**42**(24):9819-9826. DOI: 10.1016/S0032-3861(01)00509-2

[14] Ugel E, Giuliano G, Modesti M. Poly (ethylene-co-vinyl acetate)/clay nanocomposites: Effect of clay nature

and compatibilising agents on morphological thermal and mechanical properties. Soft Nanoscience Letters. 2011;**01**(04):105-119. DOI: 10.4236/snl.2011.14018

[15] Abhilash V, Rajender N, Suresh K. X-ray diffraction spectroscopy of polymer nanocomposites. In: Spectroscopy of Polymer Nanocomposites. Elsevier; 2016. pp. 410-451. DOI: 10.1016/B978-0-323-40183-8.00014-8

[16] Zanetti M, Camino G, Thomann R, Mülhaupt R. Synthesis and thermal behaviour of layered silicate-EVA nanocomposites. Polymer. 2001;**42**(10):4501-4507. DOI: 10.1016/S0032-3861(00)00775-8

[17] Bagheri S, Shameli K, Abd Hamid SB. Synthesis and characterization of anatase titanium dioxide nanoparticles using egg white solution via sol-gel method. Journal of Chemistry. 2013;**2013**:5. DOI: 10.1155/2013/848205

[18] Rimez B, Rahier H, Van Assche G, Artoos T, Biesemans M, Van Mele B. The thermal degradation of poly(vinyl acetate) and poly(ethylene-co-vinyl acetate), part I: Experimental study of the degradation mechanism. Polymer Degradation and Stability. 2008;**93**(4):800-810. DOI: 10.1016/j.polymdegradstab.2008.01.010

[19] Rimez B, Rahier H, Van Assche G, Artoos T, Van Mele B. The thermal degradation of poly(vinyl acetate) and poly(ethylene-co-vinyl acetate), part II: Modelling the degradation kinetics. Polymer Degradation and Stability. 2008;**93**(6):1222-1230. DOI: 10.1016/j.polymdegradstab.2008.01.021

[20] Pramanik M, Srivastava SK, Samantaray BK, Bhowmick AK. EVA/clay nanocomposite by solution blending: Effect of aluminosilicate layers on mechanical and thermal properties. Macromolecular Research. 2003;**11**(4):260-266. DOI: 10.1007/BF03218362

[21] Barsoum MW. Fundamentals of Ceramics. 2003. DOI: 10.1887/0750309024

[22] Davies JT. Dielectric Behaviour and Structure. New York: McGraw-Hill Book Company, Inc; 1955. pp. 199-200. DOI: 10.1016/0009-2509(56)80047-X

[23] Anastasiadis SH, Karatasos K, Vlachos G, Manias E, Giannelis EP. Nanoscopic-confinement effects on local dynamics. Physical Review Letters. 2000;**84**(5):915-918. DOI: 10.1103/PhysRevLett.84.915

[24] Wang H-W, Dong R-X, Liu C-L, Chang H-Y. Effect of clay on properties of polyimide-clay nanocomposites. Journal of Applied Polymer Science. 2007;**104**(1):318-324. DOI: 10.1002/app.25740

[25] Bershtein VA et al. Polycyanurate-organically modified montmorillonite nanocomposites: Structure-dynamics-properties relationships. Journal of Macromolecular Science, Part B Physics. 2008;**47**(3):555-575. DOI: 10.1080/00222340801955545

[26] Sengwa RJ, Sankhla S, Choudhary S. Dielectric characterization of solution intercalation and melt intercalation polyvinyl alcohol-poly(vinyl pyrrolidone) blend-montmorillonite clay nanocomposite films. Indian Journal of Pure and Applied Physics. 2010;**48**(3):196-204. DOI: 10.1016/j.compscitech.2010.06.003

[27] Sengwa RJ, Choudhary S, Sankhla S. Dielectric properties of montmorillonite clay filled poly(vinyl alcohol)/poly(ethylene oxide) blend nanocomposites. Composites Science and Technology. 2010;**70**(11):1621-1627. DOI: 10.1016/j.compscitech.2010.06.003

[28] Sengwa RJ, Choudhary S. Investigation of correlation between dielectric parameters and nanostructures in aqueous solution grown poly(vinyl alcohol)-montmorillonite clay nanocomposites by dielectric relaxation. Spectroscopy. 2010. DOI: 10.3144/expresspolymlett.2010.70

[29] Zhang LD, Zhang HF, Wang GZ, Mo CM, Zhang Y. Dielectric behaviour of nano-TiO_2 bulks. Physica Status Solidi. 1996;**157**(2):483-491. DOI: 10.1002/pssa.2211570232

[30] Nelson JK, Fothergill JC. Internal charge behaviour of nanocomposites. Nanotechnology. 2004;**15**(5):586-595. DOI: 10.1088/0957-4484/15/5/032

[31] Brosseau C, Achour ME. Variable-temperature measurements of the dielectric relaxation in carbon black loaded epoxy composites. Journal of Applied Physics. 2009;**105**(12):124102. DOI: 10.1063/1.3149702

[32] Fattoum A, Gmati F, Bohli N, Arous M, Mohamed AB. Effects of the matrix molecular weight on conductivity and dielectric relaxation in plasticized polyaniline/polymethylmethacrylathe blends. Journal of Physics D: Applied Physics. 2008;**41**(9):95407. DOI: 10.1088/0022-3727/41/9/095407

Chapter 3

A Novel Sensing Method for VOCs Using Nanoparticle-Coated Nanoporous Silicon

Selvakumar Varadarajan Subramani, Suganthi Selvakumar and Sujatha Lakshminarayanan

Abstract

Structural aspects, such as grain size, pore size, and crack-free film morphology, of porous silicon (PS), etc., play a vital role in the sensing of volatile organic compounds (VOCs). This chapter discusses a novel method for sensing of VOCs using porous silicon coated with a layer of ZnO (PS-ZnO). It was noted that the sensing ability of the PS sensor has increased due to the transconductance mechanism, as a result of the coating of ZnO over PS. Initially, porous silicon is formed by electrochemical wet etching of silicon and by electrophoretic deposition (EPD), ZnO is coated over porous silicon. An increase in the selectivity is due to the increase in surface-to-volume ratio and uniformity in the pore structures. The thickness of ZnO layer can be tuned up to 25 nm by applying a DC voltage between the copper electrode and the conductive silicon substrate immersed in a suspension of ZnO quantum dots. The influence of quantum dot concentration on the final layer thickness was studied by X-ray diffraction (XRD). The change in resistance for ethanol was found to be 12.8–16 MΩ and 8–16 MΩ for methanol.

Keywords: porous silicon, VOC sensing, metal oxide coating, electrochemical wet etching

1. Introduction

Porous silicon (PS) is a vital medium, which can be fabricated easily by electrochemical etching of silicon in HF acid solution [1]. The surface area of PS can be up to million times than that of a planar silicon surface. The melting point of silicon is lowered due to this high surface area [2]. Due to its characteristics such as high surface area and high chemical activity, it is popularly used as sensors to detect gases like NO_2, NH_3, H_2S, SO_2, ethanol, and acetone [1–3]. Compared to other semiconductor-based gas sensors (SnO_2, CuO, Cr_2O_3, and V_2O_5), PS-based gas sensors operate at comparatively lower temperatures, even at room temperature (RT) [4]. Furthermore, PS manufactured by the semiconducting material is compatible with silicon IC technology that provides the possibility to integrate the PS-based sensing element into any device [4–6]. This makes PS a promising gas-sensing material. However, the poor sensitivity and low thermal stability limit the industrial applications of PS [7].

Zinc oxide (ZnO) as an instantaneous band gap semiconductor (E_g = 3.37 eV) is used in development of gas sensors, optoelectronic devices, photovoltaic devices, and piezoelectrics [5–8]. Recent observations by Rai et al. have shown that the sensing properties of sensors depend on the morphology and surface-to-volume ratio of the sensing materials [9]. It is believed that ZnO nanostructures have high sensitivity due to their high surface-to-volume ratio and high electron mobility [5–9]. They are already being used for sensing gases such as NO_2, H_2, CO, and ethanol [5–12]. ZnO can be prepared in RT through electrophoretic deposition (EPD) of a colloidal suspension of quantum debris. EPD has been widely used to deposit particles within the micron size [13, 14]. Moreover, compared with the other processing methods, EPD is able to produce uniform deposits with high microstructural homogeneity. This helps us to control the deposit thickness and in depositing coatings on a broad range of patterns [15–17]. However, the high operating temperature and high power consumption of ZnO-based gas sensors limit its application.

The presence of a certain gas can be detected by the change in resistance produced in the sensor due to the adsorption of the analyte molecules by the PS film [18, 19]. In this chapter, we have discussed the fabrication procedure and the sensing mechanism of VOCs by PS-based gas sensor coated with ZnO (PS-ZnO) nanoparticles. The sensitivity, selectivity, and repeatability of the PS-ZnO sensor are also discussed. The primary grail of this research is to bring up a competitive impedance metric analysis of a two-point metal contact on PS-ZnO-based VOC sensor.

2. Experimental setup

2.1 Fabrication of PS-ZnO-based VOC sensor

Figure 1 shows the cross-sectional view of PS-ZnO-based VOC sensor. Since the ethanol sensing mechanism involves the formation of two layers (ZnO and PS), the following steps are followed in the manufacturing process:

1. Surface planning or silicon wafer (10 ohm-cm) cleaning

2. Electrochemical wet etching of silicon

3. Formation of permeable silicon (couple of hundred nanometer)

4. Deposition of ZnO by electrophoretic deposition

5. Making Al contact terminals utilizing screen printing system

6. Constructing the test chamber for location of ethanol, alcohol, and methanol vapor utilizing potential analyzer

The first step in the manufacturing process involves the preparation or cleaning of the silicon wafer. It is done to expel the organic and inorganic impurities that may be present on the substrate. The standard procedure followed is the RCA clean method. It is a procedure that must be performed before electrochemical wet etching.

Initially in the RCA cleaning process, organic cleaning is used to remove the natural contaminants present in thesilicon. This process is called organic clean method. The chemicals used in the organic clean process are added in the ratio of 1:1:5 corresponding to $NH_4OH/H_2O_2/H_2O$. After the initial step, ionic impurities

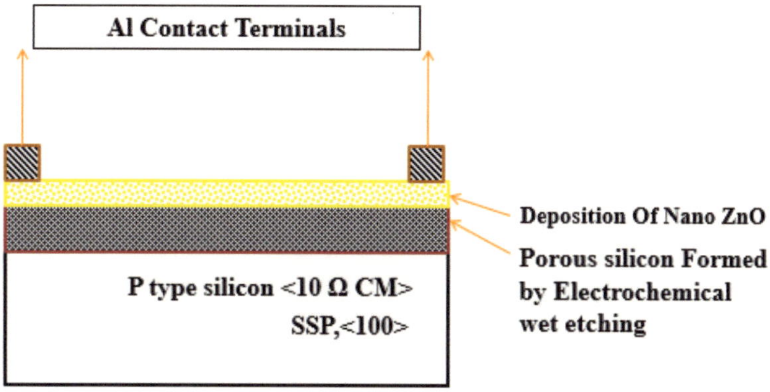

Figure 1.
Cross-sectional view of PS-ZnO-based sensor.

(ionic pollutants) are removed by a process called ionic clean method. Here the cleaning solution is in the ratio 1:1:6 corresponding to $HCl/H_2O_2/H_2O$. Once the ionic impurities are removed, the wafer is washed with deionized (DI) water, and it is later dried with the help of N_2 gas. Now the wafer is submersed in a 1:50 arrangement of HF + H_2O at 75°C usually for 10 minutes, which helps to eliminate the thin oxide layer and some of the ionic contaminants. Finally, the wafer is again flushed in a solution of deionized water and dried again by nitrogen gas.

2.2 Formation of porous silicon (PS)

The second step is the manufacturing process in the formation of the porous silicon. The initial material is a solitary crystalline silicon of p-type substrate with 1–10 ohms-cm resistivity, <100> orientation, and 525 μm thickness and is polished on one side. Initially the surface cleaning process has been done as mentioned in the previous step in the manufacturing stream. After which, the wafers have been cut into rectangular bits of 2 × 2 cm to give a uniform current distribution over the surface of the substrate. A layer of aluminum is then deposited on the bottom the silicon specimen by means of thermal evaporation.

The experimental setup for the preparation of the porous silicon can be seen in **Figure 2**. Initially, the silicon sample has been put at the base of a round and hollow Teflon cell and is properly adjusted by a tempered steel plate which acts as a support material. The cell has two electrodes as shown. The silicon wafer is considered or used as an anode, and a platinum rod placed in a HF safe material is utilized as a cathode. In our trials, the Pt mesh which acts as the cathode was taken and set vertically opposite to the anode at the highest point of the cell. The space of the cell between the two electrodes is filled with the HF-based electrolyte. The electric flow is provided by the presence of a DC steady flow source which is used to connect the two electrodes. When there is a current flowing through the circuit, the cathode excites electrolytic solution which leads to the etching of numerous small pores on the surface of the anode (Si), thus leading to the formation of porous silicon.

The porosity of PS is proportional to the current density between the electrodes and the time of the process. To attain 65% porosity, it took up to 6 minutes at a current density of 5 mA/cm^2. To attain a porosity of 75%, in our case, it took 10 minutes at a current density of 15 mA/cm^2. After the successful formation of the porous silicon, the sample specimen was tested for photoluminescence by several experiments with ethanol and isopropyl alcohol.

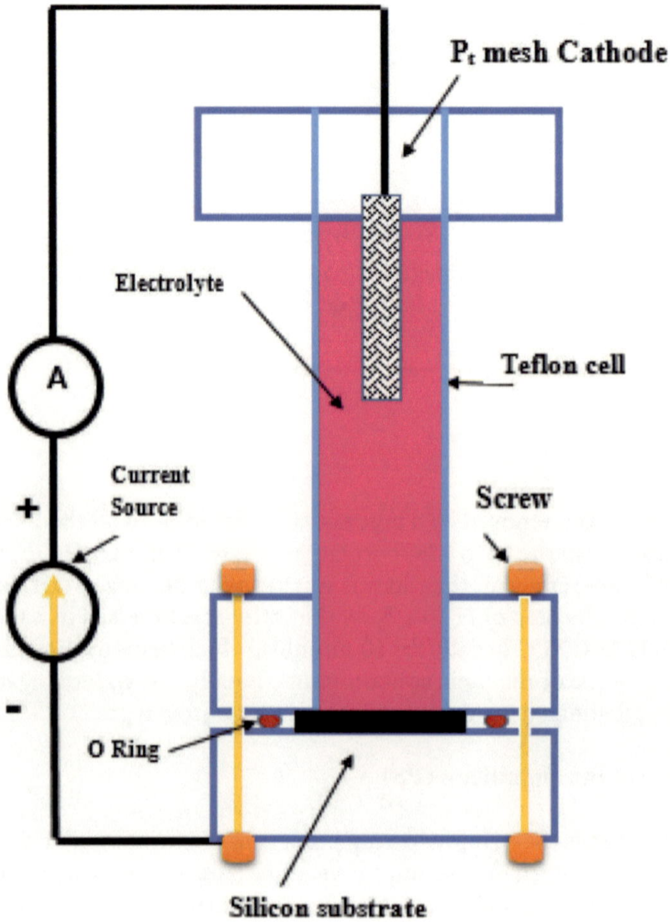

Figure 2.
Electrochemical wet etching of silicon.

From our observations described in the **Table 1**, we concluded that the photoluminescence property could be checked by the UV detection method. The UV photo illumination of porous silicon demonstrates that the reflected UV photography is excited by UV-transmitting LEDs or devices (radiation sources) or by solid daylight. A UV-transmitting channel is placed on the focal point, which enables bright light to pass and collects all visible and infrared light.

2.3 Selection of sensing materials

For optimal sensing performance, the sensing material to be selected should be carefully considered. The material selection is based on the operating temperature, range of detection, response time, sensing form, sensing parameters, and processing compatibility of the target VOC. Three materials—SnO_2, TiO_2, and ZnO—satisfy all the above parameters for the detection of VOC. However, among these three, ZnO operates at room temperature, which means that there is no need for a heating element. It also has a faster response time of less than a second and is silicon processing compatible. On the other hand, TiO_2 has a slow response time of about 3 minutes and operates only at 200–400°C. In addition, SnO_2 has a narrow range of detection and slower response time. The selectivity of SnO_2 is also less. Due to these reasons, we select ZnO as the sensing material.

DOI: http://dx.doi.org/10.5772/intechopen.86105

S. No	Name	HF/ethanol	HF/IP	HF/IP	HF/IP
1.	P-type silicon	1	1	1	1
2.	Area	2.009 cm^2	2.009 cm^2	2.009 cm^2	2.009 cm^2
3.	Current density	5 mA/cm^2	5 mA/cm^2	7.5 mA/cm^2	10 mA/cm^2
4.	Current across anode and cathode	8 mA	10 mA	15 mA	20 mA
5.	Electrolyte	(3:7 (10 ml)	1:2 (10 ml)	1:2 (10 ml)	1:1(10 ml)
6.	Pt mesh cathode	1	1	1	1
7.	O ring	1	1	1	1
8	After experiment		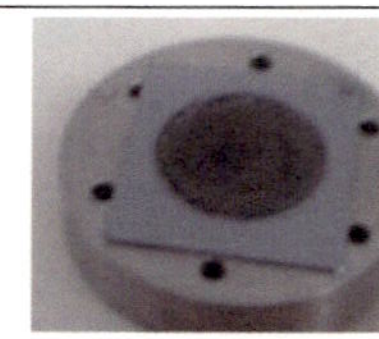	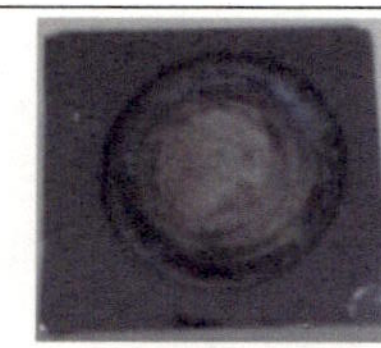	
9	Photo luminance property checked by UV detection	No photo luminance	No photo luminance		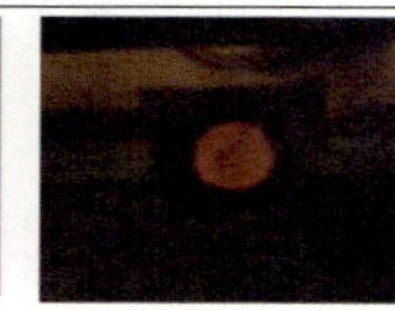

Table 1.
Different experimental results with photoluminescence.

2.4 Electrophoretic deposition of ZnO on PS

After the successful manufacturing of the porous silicon specimen, a thin layer of ZnO had to be deposited on the surface of the substrate by means of electrophoretic deposition (EPD). The mechanism of the EPD process is illustrated in **Figure 3**. A glass vessel was used for this experiment. The PS material to be coated was taken as the anode, and we utilized an aluminum rod as a cathode. The two electrodes were submerged in a solution of isopropanol which had numerous ZnO molecules suspended at the bottom of it. The apparatus had been equipped with a magnetic stirrer placed below the glass vessel which was used to stir the solution at high speeds. A DC voltage had been applied between the two electrodes which caused an electrochemical reaction between the two submerged electrodes. Initially, the suspended particles in the isopropanol were released due to the magnetic stirring process. Once the ZnO molecules start floating in the solution, the voltage caused the aluminum anode to repel the colloidal ZnO molecules, and the PS attracted them, causing a migration of particles from the positive to the negative anode, thus creating a thin layer of ZnO coating on it. After deposition, the samples were rinsed with deionized water and were air-dried with N_2 gas.

2.5 Experimental setup for VOC detection

The device was successfully fabricated and was tested for VOC detection. **Figure 4** shows the experimental setup of VOC detection by using PS-ZnO. The testing phase was conducted in an enclosed testing chamber. The specimen to be examined (PS-ZnO) was placed inside, and the chamber was tightly sealed. The chamber had been equipped with an electronic temperature controller attached to it, which regulates the temperature inside the chamber with the help of a heater coil wound around the chamber. The temperature of the coil was measured as well by means of an external thermometer to compare the actual temperature with the regulatory temperature.

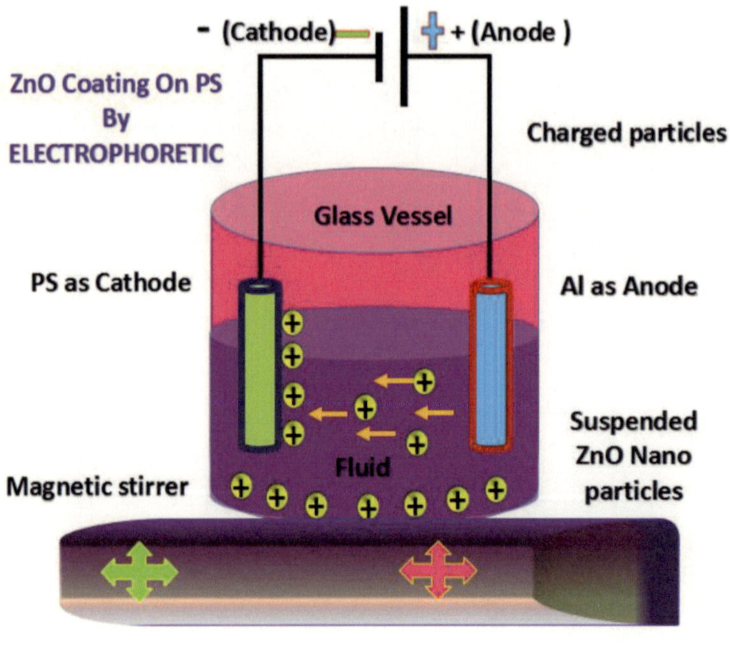

Figure 3.
Illustration of the cathodic electrophoretic deposition process.

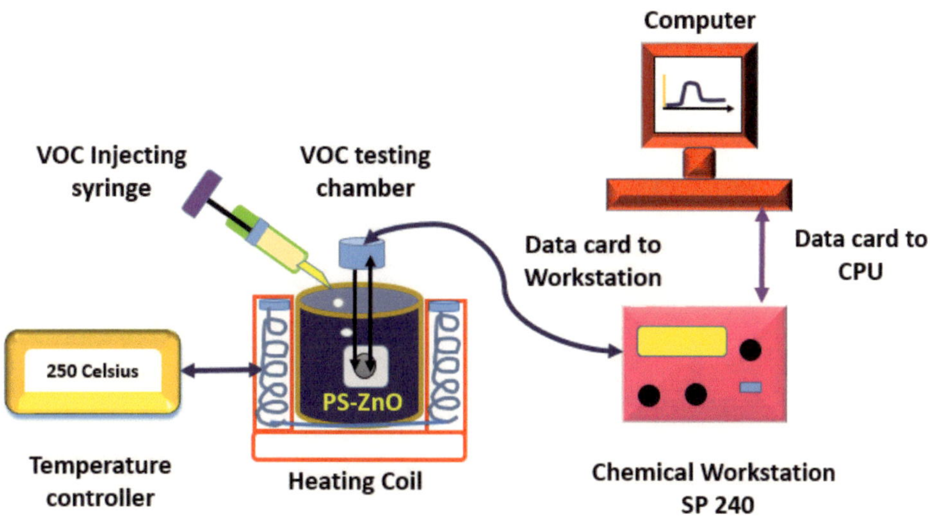

Figure 4.
VOC recording system through chemical workstation.

For our study we had set the regulated temperature in the chamber to 250°C. A data cord was then connected from the testing chamber to a chemical workstation (SP 240). The chemical workstation was then further connected to a CPU to observe readings. The VOC compound to be tested was taken in a syringe and was slowly injected into the gas testing chamber. The injection of the compound was done in a gradual manner, and the corresponding data (impedance characteristics) on the display were observed and noted.

3. Results and discussion

The pictures below in **Figure 5(a)** and **(b)** show the structure of porous silicon before and after the electrophoretic deposition of ZnO on its surface, respectively. The images were used to illustrate the basic topological differences in appearance and structure of the PS substrate between the two stages. The microscopic images of micro PS samples shown in **Figure 6** were used to study the morphology of the substrate after the addition of the ZnO layer. The bright-field microscopic images clearly show

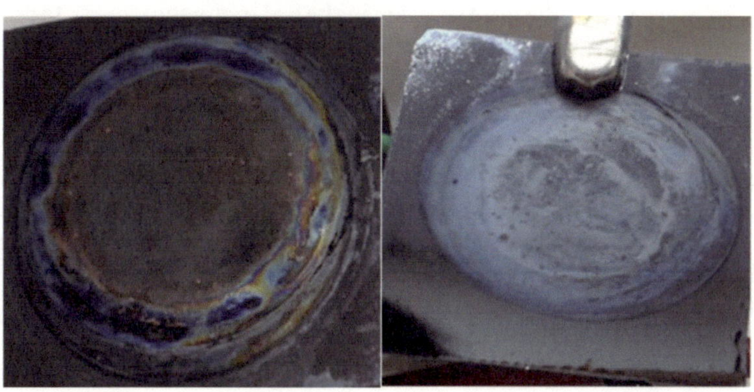

(a) (b)

Figure 5.
(a) PS before ZnO deposition by EPD; (b) PS after deposition of ZnO by EPD.

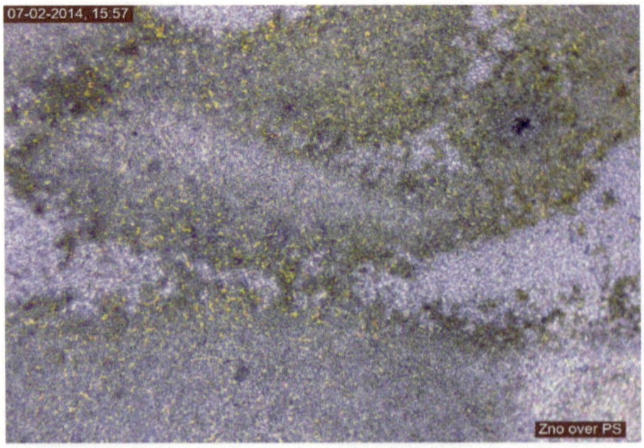

Figure 6.
Microscopic image of PS with ZnO.

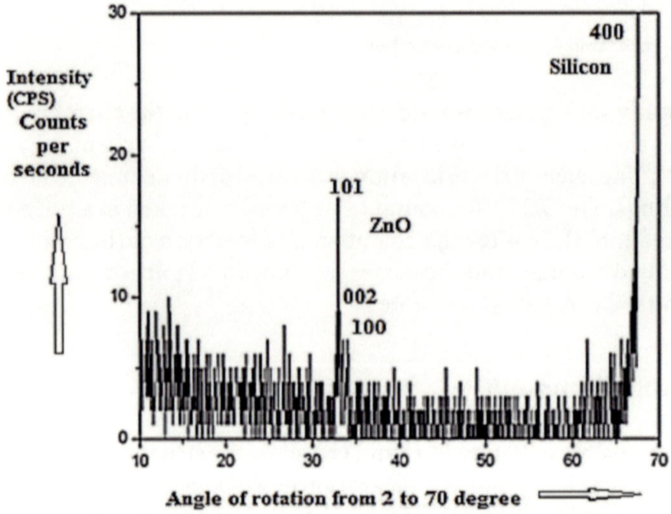

Figure 7.
XRD conformation of ZnO and silicon.

the structural changes in the obtained porous silicon after the coating of ZnO by EPD process. It is observed that the pore sizes are in the order of nanometers. In addition, the pores in micro PS are random in nature with uniform deposition of ZnO.

The coating characterization of the device was tested using spectrum diffraction analysis of both the materials. **Figure 7** shows the conformed X-ray diffraction (XRD) spectrogram of ZnO nanoparticles that are coated on the surface of the PS substrate. Crystalline form of ZnO and PS was taken for the spectrum analysis, and the intensity characteristics were studied by varying the angle of rotation between 2 and 70°. Distinctive intensity peaks of ZnO and the porous silicon material were observed at 32°90′ and 69°8′, respectively. We also observed that the intensity in counts per second (CPS) was much higher for the porous silicon (400) compared to the deposited layer of the ZnO (101).

After studying the coating characterization of ZnO, the sensitivity and the responsivity of the sensor were studied. An impedance study was done with the ZnO-coated PS sensor for certain VOCs, namely, ethanol, isopropanol, and methanol. We also observed the rate-limiting processes that occur at particular frequencies during operation of the device in the impedance spectra.

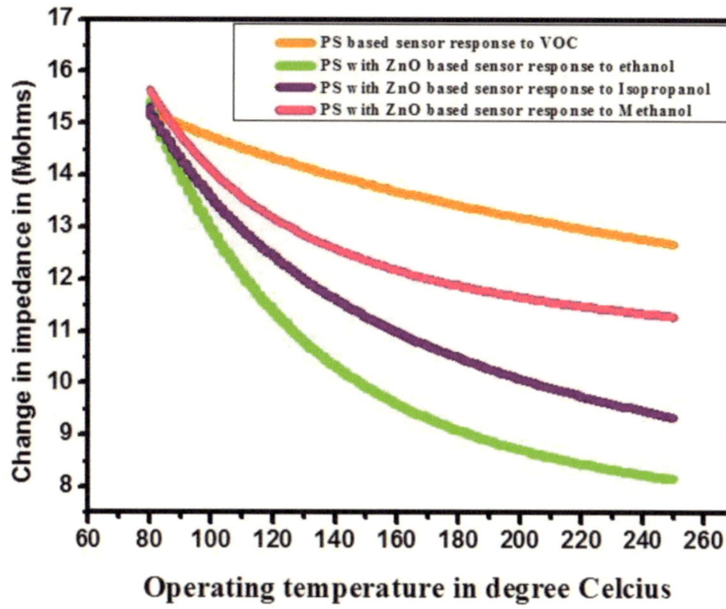

Figure 8.
Sensor response for ethanol, isopropanol, and methanol.

Figure 8 shows a typical Nyquist plot of data recorded from a PS-based sensor with an Al working electrode at room temperature. The plot also shows the results of two measurements made (resistance and temperature) when the sensor was exposed to three different VOCs: (1) ethanol, (2) methanol, and (3) isopropanol. **Figure 8** shows the measure of the change in resistance of the sensor at different temperatures ranging from 70–250°C. It was observed that the change in resistance varied from 8 to 15.8 MΩ for methanol and from 12 to 15.6 MΩ for isopropanol.

The measured change in resistance for ethanol was observed as well which varied from 12 to 15.8 MΩ at the same temperature gradient. We observed that the greatest percentage of change in the impedance measurement for the given temperature range was the highest for methanol. It is also evident that the percentage of change in resistance for the PS-based sensor without the ZnO coating is lower than the device coated with the ZnO film. The result thus proves that the sensing properties of a PS-based device are significantly increased by the addition of ZnO coating.

After observing the impedance characteristics of the device, we tested the responsivity of the sensor for different concentrations of ethanol. **Figure 9** shows the response of the ZnO-based VOC sensor for different concentrations of ethanol at a specific temperature of 250°C. We observed that as the concentration of ethanol kept increasing, the resistance of the device kept decreasing. This gave a linear increase in responsivity from 0 to 255 ppm of ethanol after which the increase in responsivity was relatively less due to saturation of ethanol level. After studying the sensitivity and responsivity of the sensor, the selectivity was studied as well.

Figure 10 summarizes the selectivity of PS toward aliphatic alcohol with respect to various VOCs at a frequency of 100 Hz. We observed that the selectivity of methanol was around 90%, while the same for ethanol and isopropanol compounds was around 50 and 60%, respectively. It shows that the response and recovery of the sensor are about 2 and 27 s, respectively. This indicates the response and recovery performance of the developed sensor toward methanol, ethanol, and isopropanol. The various quick responses and recoveries of methanol, ethanol, and isopropanol

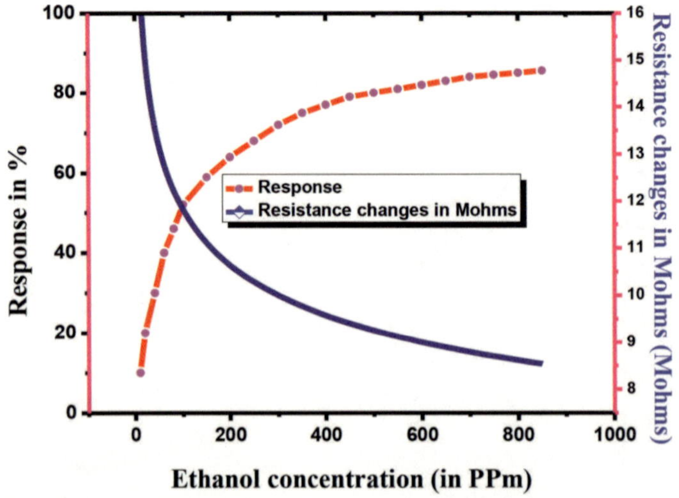

Figure 9.
Response and sensitivity for ethanol concentration.

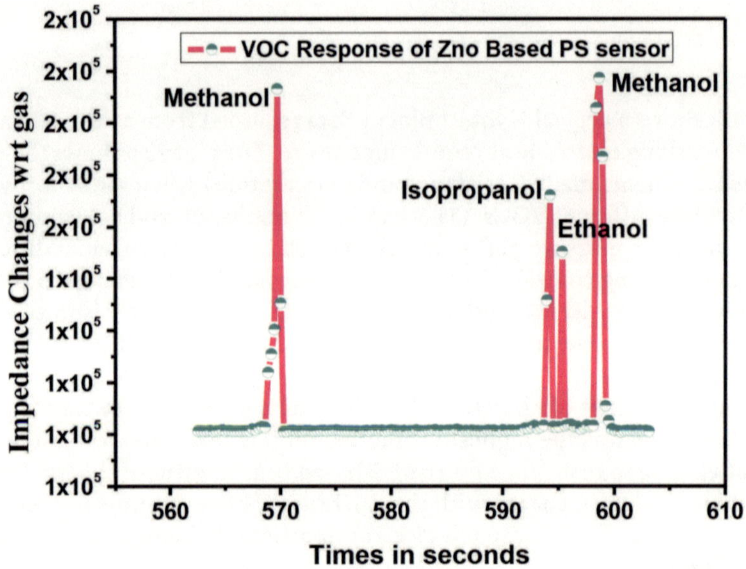

Figure 10.
Sensor repeatability test for different VOC concentrations.

clearly prove selectivity of these VOCs by measuring different impedances. The ZnO-PS-based sensor is very sensitive to methanol when compared to all other VOCs.

From **Figure 11**, it is understood that the selectivity and repeatability are one of the range of acceptable values the sensor will read, relevant to a previous measurement, when measuring a methanol, ethanol, and isopropanol, independent of the actual gas concentration. It clearly shows that the sensing action remains unchanged for repeated measurement.

In the modern world, nanostructured oxide-based VOC sensors are used in applications such as monitoring the environment, chemical process control equipment, and health regulation devices. However, the crux of the operating mechanism is still not completely understood. Therefore, a deeper understanding would be beneficial to

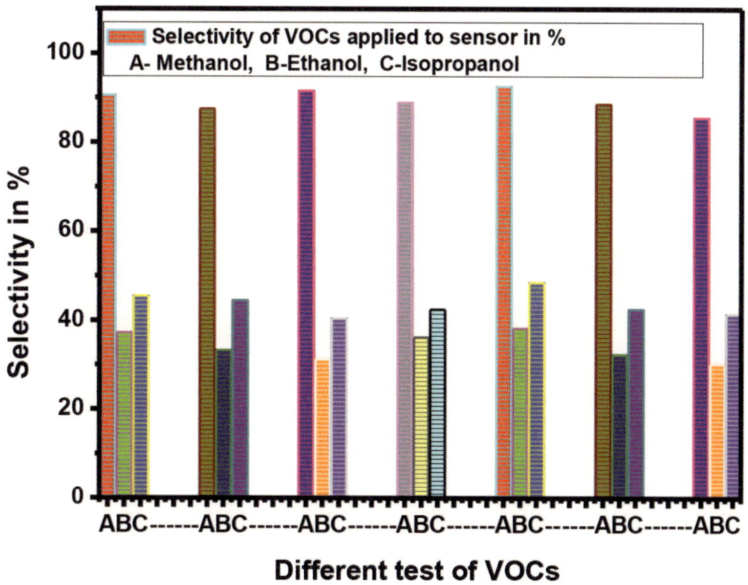

Figure 11.
Sensor selectivity for VOC concentration in repeated test.

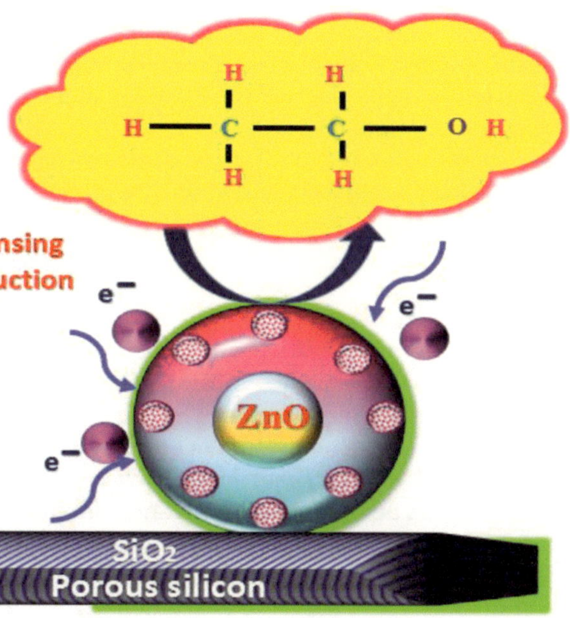

Figure 12.
VOC sensing mechanism of ZnO-based PS sensor.

the development of enhanced sensing elements which can be used for the mentioned applications. In the case of ethanol sensing, our outcomes uncover that the focused adsorption of electrons at the nanostructure surfaces between the porous silicon and the ethanol particles via reduction process (as shown in **Figure 4**) indicates the presence of ethanol at that region. Therefore, the process of ethanol fixating on the ZnO can be identified electronically, helping us discover the sensing element.

Initially, there is an interaction between the ethanol molecules and the chemically absorbed oxygen ions that are present in the sensing element (ZnO). The

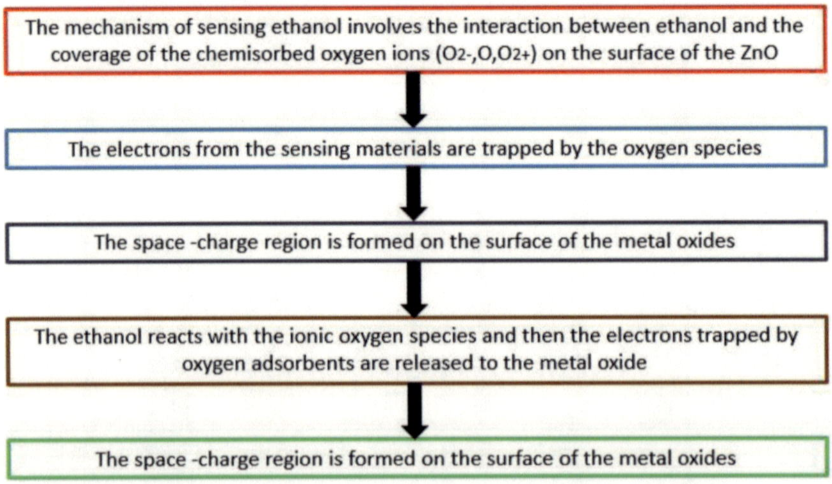

Figure 13.
VOC sensing mechanism of nanoparticle-based PS sensor.

electrons that are present in the sensing material are already trapped by these oxygen compounds or species. Due to this, a surface charge layer is formed on the surface of the metal oxide (ZnO). This in turn causes the ethanol to react with the ionic oxygen species that are present in the sensing material which leads to the release of the trapped electrons into the metal oxide. Finally, it reverts back to the space charge region on the surface of the metal oxide. **Figure 12** shows the graphical representation VOC sensing mechanism ZnO-based PS sensor; it clearly represents the reduction of ethanol (electron exchange with sensing layer) sensing mechanism of ZnO. The mechanism flow of sensing is expressed as a flow diagram in **Figure 13**.

4. Conclusion

The novel method for sensing of VOCs using a porous silicon sensor coated with a layer of ZnO has been presented. The key structural aspects have been explained in detail. The coating of ZnO over PS by EPD technique was carried out successfully. The XRD spectrography was used to successfully obtain distinct intensity peaks of 101 and 400 for ZnO and porous silicon which were found at 32.90 and 69.8, respectively. The resistance changes, which varied from 8 to 16 MΩ for methanol, 12 to 15.6 MΩ for isopropanol, and 12 to 15.8 MΩ for ethanol, were observed and studied carefully. This proved that device exhibited higher sensitivity and selectivity at the temperature range between 70 and 250°C for PS sensor coated with ZnO nanoparticles. We also concluded that the device works best for the sensing of ethanol as it showed high repeatability of up to 80% at saturated levels and has high selectivity of about 90% for methanol.

Acknowledgements

We would like to acknowledge the INUP, IISc, Bangalore, and the NPMASS Program for the establishment of National MEMS Design Centre (NMDC) at Rajalakshmi Engineering College.

Author details

Selvakumar Varadarajan Subramani[1*] Suganthi Selvakumar[1] and
Sujatha Lakshminarayanan[1,2]

1 Department of Electronics and Communication Engineering, Rajalakshmi Engineering college, Chennai, India

2 Centre of Excellence in MEMS and Microfluidics (CEMM), Rajalakshmi Engineering College, Chennai, India

*Address all correspondence to: selvakumar.vs@rajalakshmi.edu.in

References

[1] Razi F, Rahimi F, Irajizad A. Fourier transform infrared spectroscopy and scanning tunnelling spectroscopy of porous silicon in the presence of methanol. Sensors and Actuators B: Chemical. 2008;**132**(1):40-44

[2] Northen TR, Woo H-K, Northen MT, Nordström A, Uritboonthail W, Turner KL, et al. High surface area of porous silicon drives desorption of intact molecules. Journal of the American Society for Mass Spectrometry. 2007;**18**(11):1945-1949

[3] Ozdemir S, Gole JL. A phosphine detection matrix using nanostructure modified porous silicon gas sensors. Sensors and Actuators B: Chemical. 2010;**151**(1):274-280

[4] Ali NK, Hashim MR, Aziz AA. Effects of surface passivation in porous silicon as H_2 gas sensor. Solid-State Electronics. 2008;**52**(7):1071-1074

[5] Öztürk S, Kılınç N, altin NŢ, Öztürk ZZ. A comparative study on the NO_2 gas sensing properties of ZnO thin films, nanowires and nanorods. Thin Solid Films. 2001;**520**(3):932-938

[6] Balucani M, Nenzi P, Chubenko E, Klyshko A, Bondarenko V. Electrochemical and hydrothermal deposition of ZnO on silicon: From continuous films to nanocrystals. Journal of Nanoparticle Research. 2011;**13**:5985-5997

[7] Kanungo J, Saha H, Basu S. Pd sensitized porous silicon hydrogensensor-influence of ZnO thin film. Sensors and Actuators B: Chemical. 2010;**147**(1):128-136

[8] Chen YJ, Zhu C-L, Xiao G. Ethanol sensing characteristics of ambient temperature Nano chemically synthesized ZnO nanotubes. Sensors and Actuators B: Chemical. 2008;**129**(2):639-642

[9] Rai P, Raj S, Ko KJ, Park KK, Yu YT. Synthesis of flower-like ZnO microstructures for gas sensor applications. Sensors and Actuators B: Chemical. 2013;**178**(1):107-112

[10] Bai S, Sun C, Guo T, Luo R, Lin Y, Chen A, et al. Low temperature electrochemical deposition of nanoporous ZnO thin films as novel NO_2 sensors. Electrochimica Acta. 2013;**90**:530-534

[11] Bie LJ, Yan XN, Yin J, Duan YQ, Yuan ZH. Nano pillar ZnO gas sensor for hydrogen and ethanol. Sensors and Actuators B: Chemical. 2007;**126**(1):604-608

[12] Wan Q, Li QH, Chen YJ, Wang TH, He XL, Li JP, et al. Fabrication and ethanol sensing characteristics of ZnO nanowire gas sensors. Applied Physics Letters. 2004;**84**(18):3654

[13] Sarkar P, Nicholson PS. Electrophoretic deposition (EPD): Mechanism, kinetics and application to ceramics. Journal of the American Ceramic Society. 1996;**79**(1):1987-2002

[14] Boccaccini AR, Zhitomirsky I. Application of electrophoretic and electrolytic deposition techniques in ceramics processing. Current Opinion in Solid State & Materials Science. 2002;**6**(3):251-260

[15] Van der Biest OO, Vandeperre LJ. Electrophoretic deposition of materials. Annual Review of Materials Science. 1999;**29**:327-352

[16] Babsu S, Wang Y-h, Ghanshyam C, Kapur P. Fast response time alcohol gas sensor using nanocrystalline F-doped SnO_2 films derived via sol–gel method. Bulletin of Materials Science. 2013;**36**(4):521-533

[17] Jawad MJ, hashim MRR, Ali NK. Improvement of Al metal contact on porous silicon. Progress of Physics Research in Malaysia, AIP Conference Proceedings. 2010;**1250**(1):73-76

[18] Selvakumar VS, Sujatha L, Sundar R. Design analysis and electrical characteristics of porous silicon structure for sensor applications. Sensor Letters. 2018;**16**(4):1-6

[19] Selvakumar VS, Sujatha L, Sundar R. A novel MEMS microheater based alcohol gas sensor using nanoparticles. Journal of Semiconductor Technology and Science. 2018;**18**(4):445-453

Chapter 4

Optical Tweezers in Biotechnology

Zhiyong Gong and Yuchao Li

Abstract

Three-dimensional optical manipulation of microparticles, cells, and biomolecules in a noncontact and noninvasive manner is crucial for biophotonic, nanophotonic, and biomedical fields. Optical tweezers, as a standard optical manipulation technique, have some limitations in precise manipulation of micro-objects in microfluidics and in vivo because of their bulky lens system and limited penetration depth. Moreover, when applied for trapping nanoscale objects, especially with sizes smaller than 100 nm, the strength of optical tweezers becomes significantly weak due to the diffraction limit of light. The emerging near-field methods, such as plasmon tweezers and photonic crystal resonators, have enabled surpassing of the diffraction limit. However, these methods msay lead to local heating effects that will damage the biological specimens and reduce the trapping stability. Furthermore, the available near-field techniques rely on complex nanostructures fixed on substrates, which are usually used for 2D manipulation. The optical tweezers are of great potential for the applications including nanostructure assembly, cancer cell sorting, targeted drug delivery, single-molecule studies, and biosensing.

Keywords: optical tweezers, optical force, optical manipulation, biophotonics, biotechnology

1. Introduction

As early as the 1970s, Maxwell revealed that electromagnetic waves can carry momentum in his famous electromagnetic field theory. When electromagnetic waves are applied to objects, they will transmit momentum. Therefore, electromagnetic waves can exert force on objects, and then the concept of electromagnetic force is proposed [1]. Until the beginning of the twentieth century, Einstein proposed the concept of photonic quantum, which believes that light is composed of a group of photons with both mass and momentum. When light is irradiated on the surface of the object, it can cause changes in the photon momentum to produce radiation pressure on the object due to the scattering and absorption of light. Subsequently, Lebedev, Nichol, and Hull first demonstrated the existence of radiation pressure experimentally. The experiment used arc lamps and torsion scales to observe the effect of light in the macroscopic physical world. However, the light produced by the arc lamp is very weak and difficult to practically apply. Until 1960, the invention of the laser provided a high-intensity optical source for studying optical force, which greatly promoted the application of optical manipulation. Arthur Ashkin, a scientist at Bell Experiments in the United States, first used the radiation pressure generated by the laser beam to push tiny particles in the liquid environment [2] and then used two opposing laser beams to capture microparticles and even atoms. However, the experimental setup used in the dual-beam capture method is too complex and can only limit microparticles in a two-dimensional plane. Scientists hope to use a single laser to achieve three-dimensional trapping of microparticles. To this end, in 1986, Ashkin et al. used a high-numerical-aperture objective to focus a single laser to trap microparticles and named the technology "single beam gradient force trap" [3]. A year later, Ashkin et al. continued to improve this technology and achieved optical trapping and manipulation of tiny bacteria and viruses. They officially named the technology "optical tweezers" [3]. Compared with traditional macro-mechanical tweezers, the optical tweezers have the advantages of noncontact and no damage and can perform high-precision manipulation of microscopic particles. Therefore, since the birth of the optical tweezer technology, it has played an important role in the fields of biomedicine and physical chemistry.

1.1 Traditional optical tweezers

1.1.1 Basic principles

The core component of the traditional optical tweezers is a highly focused beam, as shown in **Figure 1a** [4]. When the incident laser (usually a near-infrared laser with a wavelength of 1064 nm) is focused by a high-numerical-aperture objective lens, the microparticles in the liquid environment will be exposed to optical force near the focus. This force is derived from the momentum transfer effect between light and particles. Specifically, the optical forces are divided into two components: one component along the direction of the optical gradient, called the optical gradient force, which is caused by the microparticles being in a nonuniform optical field, and the optical gradient force, which drives the particles to the area where the optical intensity is greatest; another component along the direction of optical propagation, called optical scattering force, is caused by the scattering and absorption of particles, and the optical scattering force causes the microparticles to move along the direction of optical propagation. By modulating the focused beam, the magnitude of two forces can be varied to achieve different functions such as capture, acceleration, and rotation of the microparticles. For traditional optical tweezers to construct a stable trap, it is necessary to focus the incident laser with a high-numerical-aperture (generally NA = 1.0~1.4) objective lens. The resulting

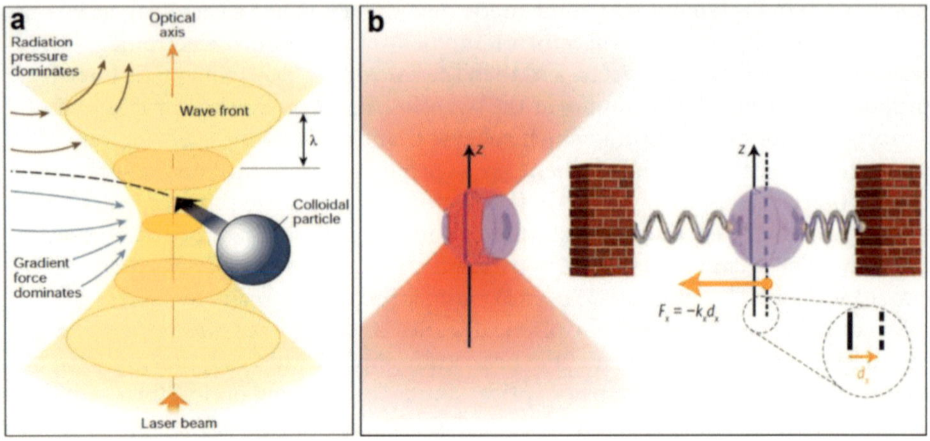

Figure 1.
Schematic diagram of the traditional optical tweezers. (a) A single microparticle is trapped to the focused spot of a laser beam by gradient force and scattering force to [4]. (b) A simple harmonic oscillator model for the optical trapping of the microparticle.

optical gradient force is greater than the optical scattering force, so the microparticles or the cells can be stably trapped in the focus of light [5].

The model in which the object is trapped by the optical tweezers can be equivalent to a simple oscillator, as shown in **Figure 1b**. The magnitude of the object's received optical force (F) is proportional to the object's distance from the focus (d), which is

$$F = -kd \tag{1}$$

where the constant k represents the spring constant of the spring oscillator and the strength of the trap. Therefore, when we know the motion of an object in a trap, the magnitude of the optical force can be calculated by Formula 1. However, in the more general case, we want to quantitatively analyze the optical force when the unknown object motions and then other optical theories are needed. The theoretical analysis of optical tweezers needs to be determined according to the size of the object, specifically divided into three cases: first, when the radius (R) of the particle is much larger than the wavelength (λ) of the incident light, then a simple geometric optical method can be used to analyze the force of the object; second, if the size of the particle is much smaller than the wavelength of the incident light, the particle can be equivalent to the dipole in the electric field, and a dipole approximation model is needed; and third, if the size of the particles is close to the wavelength of the light, the situation becomes complicated, and the Maxwell equation is needed to solve the problem.

We first analyze the Rayleigh nanoparticle ($R << \lambda$). At this time, the nanoparticle can be regarded as a dipole in a nonuniform electromagnetic field, and the optical gradient force (F_{grad}) of the dipole in the electromagnetic field can be expressed as

$$\boldsymbol{F}_{\text{grad}} = \frac{1}{2}|\alpha|\ \nabla\langle \boldsymbol{E}^2\rangle, \tag{2}$$

where α is the polarizability of the dipole, $\boldsymbol{E}$ is the electric field, parentheses indicate the time average, and */E2/* is proportional to the intensity of the electromagnetic field. It can be seen from Formula 2 that the direction of the optical gradient force $\boldsymbol{F}_{grad}$ is along the direction of the optical intensity gradient. Thus, for a highly concentrated beam, the particles are drawn to the focus of the spot. Here, the polarizability α is a crucial parameter that directly determines the intensity of the interaction of light with object. For spherical nanoparticles, α can be expressed as [6]

$$\alpha = \alpha_0 \left(1 - \frac{ik^3\alpha_0}{6\pi\varepsilon_0}\right)^{-1}, \tag{3}$$

where $k = 2\pi n/\lambda$ is the scalar of the incident light wave vector, ε is the dielectric constant of the particle, ε_0 is the dielectric constant of the vacuum, α_0 is the quasi-static polarizability of the nanoparticle, and α_0 can be given by the Clausius-Mossotti relation [6]:

$$\alpha_0 = 4\pi\varepsilon_0 R^3 \frac{\varepsilon - 1}{\varepsilon + 2}, \tag{4}$$

The radiation pressure ($\boldsymbol{F}_{rad}$) is produced by the scattering and absorption of light by the surface of the particles, which can be expressed as [7]

$$\boldsymbol{F}_{\text{rad}} = \frac{n\langle \boldsymbol{P}\rangle}{c}\sigma, \tag{5}$$

where n is the refractive index of the surrounding environment, c is the speed of light in the vacuum, and $\langle \mathbf{P} \rangle$ is the time-averaged Poynting vector, which can be expressed as

$$\langle \boldsymbol{P} \rangle = \frac{1}{2} Re(\boldsymbol{E} \times \boldsymbol{H}^*). \tag{6}$$

The σ in Formula 6 reflects the characteristics of the nanoparticle, which indicates the extinction cross section of the nanoparticle, including the scattering cross section (σ_{scat}) and the absorption cross section (σ_{abs}), and σ is determined by the following formula [8]:

$$\sigma = \sigma_{scat} + \sigma_{abs} = \frac{k^4 |\alpha|^2}{4\pi} + k\alpha'', \tag{7}$$

where α'' is the imaginary part of the particle polarizability α, which represents the absorption of light by the particles. For transparent media particles, this term is approximately equal to zero and can be ignored. It can be seen from Formula 5 that the direction of the optical scattering force coincides with the direction of the glass booth vector, that is, the direction in which the optical scattering force propagates along the light. When $\boldsymbol{F}_{grad} > \boldsymbol{F}_{rad}$, the trapping of particles can be achieved.

The dipole approximation model is only applicable to spherical nanoparticles. When the shape of the captured object is irregular or the size is the same magnitude as the wavelength, it needs to be solved from the most basic Maxwell equations using simulation software. This method is based on the Maxwell stress tensor integral of the surface S of the object, as defined below:

$$F_O = \oint_S (\langle T_M \rangle \cdot n) dS, \tag{8}$$

where $\boldsymbol{n}$ is the normal vector of the surface of the object and$\langle \mathbf{TM} \rangle$ is the time-averaged Maxwell stress tensor. The expression is

$$\langle \boldsymbol{T}_M \rangle = \frac{1}{2} Re\left[\varepsilon \boldsymbol{E}\boldsymbol{E}^* + \mu \boldsymbol{H}\boldsymbol{H}^* - \frac{1}{2}\left(\varepsilon |\boldsymbol{E}|^2 + \mu |\boldsymbol{H}|^2\right)\boldsymbol{I} \right], \tag{9}$$

where E and H are the electric field vector and the magnetic flux vector in the electromagnetic field, E^* and H^* are complex conjugates, $\boldsymbol{I}$ is an isotropic tensor, and ε and μ represent the dielectric constant and magnetic permeability, respectively. After calculating the optical force, the torque of the object can also be calculated by the following formula:

$$\boldsymbol{T} = \int \boldsymbol{rp} \times \boldsymbol{dFp}, \tag{10}$$

where dFp represents the unit force at the point of action p and rp is the position vector from the center of the object to the point of action p.

1.1.2 Applications of the optical tweezers

Professor Ashkin, the pioneer of optical tweezers, predicted that optical tweezers as the manipulation technology of tiny particles will be widely used in the research of molecular biology, cell biology, and mesoscopic physics, especially to promote the development of many interdisciplinary subjects [9]. As an example, we will introduce some of the applications of the optical tweezers in the following aspects:

1.1.2.1 Capture, separation, and assembly of microparticles and cells

The invention of optical tweezers was used to capture and manipulate tiny particles such as polystyrene microspheres, biological cells, viruses, and bacteria [12]. By capturing these tiny particles, the Brownian motion of particles can be overcome and fixed in the field of the microscope for the researcher to observe and detect. When the particles are stably captured, they can be moved to a specific position and arranged in a regular pattern, which is applied to the ordered assembly of particles and cell arrays (as shown in **Figure 2a**), giving it a specific function. Further, by measuring the mechanical properties of particles and cell array, the interaction between the particles or cells can be studied. In addition, since different types of particles and cells are affected by the magnitude and direction of optical force, separation and screening of particles and cells can be achieved. With the maturity of optical tweezer technology, the system of optical tweezers is gradually combined with Raman technology, fluorescence technology [13], confocal technology, and femtosecond laser technology and achieves real-time detection of captured targets, which will enrich the applications of optical tweezers in cell biology and colloidal physics.

1.1.2.2 Study of optical tweezers and single molecules

The optical technology has a high mechanical resolution (10^{-12}–10^{-15} N), which is sufficient for the study of individual biomacromolecules. For example, the basic laws of life movement are explained by measuring the physical forces such as the tiny force of biological single molecule and the motion step size. Optical tweezer technology has become an indispensable tool for quantitatively studying life processes and transforming life activities. Since the diameter of biomolecules is generally between 1 and 10 nanometers, the optical tweezer system cannot directly observe and manipulate. In order to see a single molecule, it is necessary to combine fluorescence imaging technology; in order to manipulate a single molecule, it is necessary to connect the molecule to the microsphere and indirectly manipulate and measure by using the small microsphere as the "handle" of the manipulation. For example, the

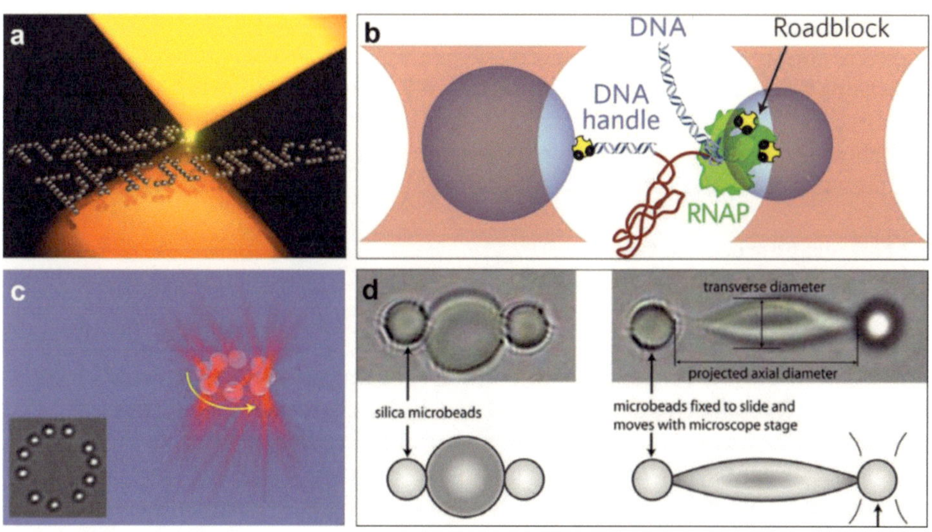

Figure 2.
Several application examples of traditional optical tweezers. (a) Order and assemble microparticles and cells. (b) Study the interaction of nucleic acid molecules using micron media balls as handles [10]. (c) Rotating the microspheres using a vortex beam [11]. (d) Stretching human red blood cells using a micron media ball as a handle.

two ends of the DNA molecular chain are, respectively, connected to two microspheres, and the microspheres are manipulated by a double-beam tweezers to stretch the DNA molecular chain and measure its elastic properties (as shown in **Figure 2b**) [10]. By rotating the two microspheres in the opposite direction, the binding force of the DNA molecular chain can be calculated. Using similar methods, researchers can also study the properties of various biomacromolecules: RNA transcription, kinesin movement, the role of polymerases, etc. These are the basic processes of life activities. Its high-precision measurement can reveal the basic laws of life activities and lay the foundation for the research and application of biomedicine.

1.1.2.3 Optical rotator

The optical rotator is a branch of the optical tweezers that not only captures the microparticles but also allows the angular rotation of the microparticles as shown in **Figure 2c** [11]. This technique is based on the moment applied by the angular momentum of the light to the object. In order to achieve the rotation of the particles, the optical rotator requires a special beam of angular momentum, such as a Laguerre-Gauss beam [14]. Rotating particles or cells are used in many fields, such as rotating a tiny mechanical motor in a liquid environment to control the movement of local water flow. In addition, by rotating living cells, it can be imaged at various angles, which is beneficial to observe the full three-dimensional appearance of cells.

1.1.2.4 Optical stretchers

Stretching cells can study the elasticity of cell membranes, and the elasticity of cell membranes is closely related to many cellular diseases and can be used to reflect the activity of cells and even the health of the human body. There are many optical stretching methods based on optical tweezers, such as direct stretching of double-beam tweezers, stretching by microsphere handle, time-division multiplexed stretching, and so on. The method based on the microsphere handle-stretching method is more commonly used because of the high measurement precision. The method is shown in **Figure 2d**: two microspheres are adhered to the cell surface by chemical coupling, and then the microspheres are controlled to move in opposite directions by the tweezers. At this time, the cell membrane is stretched by shearing force. By recording the shape variables of the cells and measuring the force of stretching the microspheres, physical parameters such as the elastic modulus of the cell membrane can be calculated.

1.2 Holographic optical tweezers

1.2.1 Basic principles

Traditional optical tweezers based on single beam can only capture and manipulate one or a few particles at a time. However, researchers want to improve the efficiency of capture, such as controlling multiple particles at the same time. Based on this goal, scientists invented holographic optical tweezers. The core component of holographic optical tweezers is a hologram element: an interference pattern formed by recording the object light and reference light through the film. The wave front can be adjusted by holographic elements to construct a light field with a specific function. The holographic optical tweezers were firstly invented in 1998 by Professor Grier of the University of Chicago and his collaborators [15]. They used a holographic element (diffraction grating) to split the collimated single laser beam into multiple independent beams, and then an array of grating is formed by focusing the lens to capture a large number of microparticles. The earliest holographic elements were prepared by coherent-optical interferometry, but the holographic elements obtained by this method have low diffraction efficiency and poor versatility, and thus this method has not been widely used. In order to improve diffraction efficiency and applicability, conventional holographic elements are often composed of spatial light modulators. The spatial light modulators include liquid crystal spatial light modulators, acousto-optic modulators, and digital microlens arrays. The spatial light modulator is controlled by a computer, and each focused beam can be individually controlled by changing the hologram element so that the formed trap well can be dynamically changed. Such holographic optical tweezers not only capture a plurality of microparticles at the same time but also control the movement of each microparticle to be arranged in different shapes, thereby achieving ordered assembly of the microparticles.

1.2.2 Applications of the holographic optical tweezers

As an emerging optical technology, holographic optical tweezers can trap and manipulate a large number of particles, showing great application prospects in the fields of particle assembly and construction of three-dimensional cell microstructure (**Figure 3**). For example, Glen R. Kirkham et al. of the United Kingdom used holographic optical tweezers to assemble one-, two-, and three-dimensional embryonic stem cell array structures (as shown in **Figure 4**) to provide a new means to study the directed differentiation of stem cells [16]. Moreover, Jesacher and his colleagues from Austria regulated the amplitude and phase of the incident light field through a liquid crystal spatial light modulator, which not only realized trapping potential wells of special shapes such as line, cross, circle, and rectangle but also controlled the microparticle movement along a specific path. In addition, holographic optical tweezer technology can also produce beams with special modes, such as Bessel beams, Laguerre-Gauss beams, and Airy beams [18]. These special-mode beams have peculiar phase distribution and propagation characteristics and can generate trapped potential wells with special functions, such as rotating particles with a Laguerre-Gauss beam, which can be used to construct micro- and nano-motors and study the transfer of orbital angular momentum; Airy beam or Bessel beam can be used to transport particles for sorting different types of particles and cells.

1.3 Fiber-based optical tweezers

1.3.1 Basic principles

Due to the low integration of conventional optical tweezer systems, it is difficult to manipulate particles located in a narrow position, such as particles inside a microfluidic channel or red blood cells in a blood vessel. The newly developed fiber-based optical tweezers are promising candidates because of its compact structure and flexible operation, which can overcome the problems of traditional optical tweezers [19]. Fiber-based optical tweezers use the output light from the end face of the fiber to achieve particle capture and manipulation, as shown in **Figure 5a**. When the laser beam passes into the fiber, it converges through the end of fiber and form a highly focused beam. The microparticles located near the tip of the fiber will be captured by the longitudinal gradient force onto the optical axis of the fiber and then captured by the lateral gradient force at the focus of the emitted light or move along the optical axis under the action of optical scattering force. For

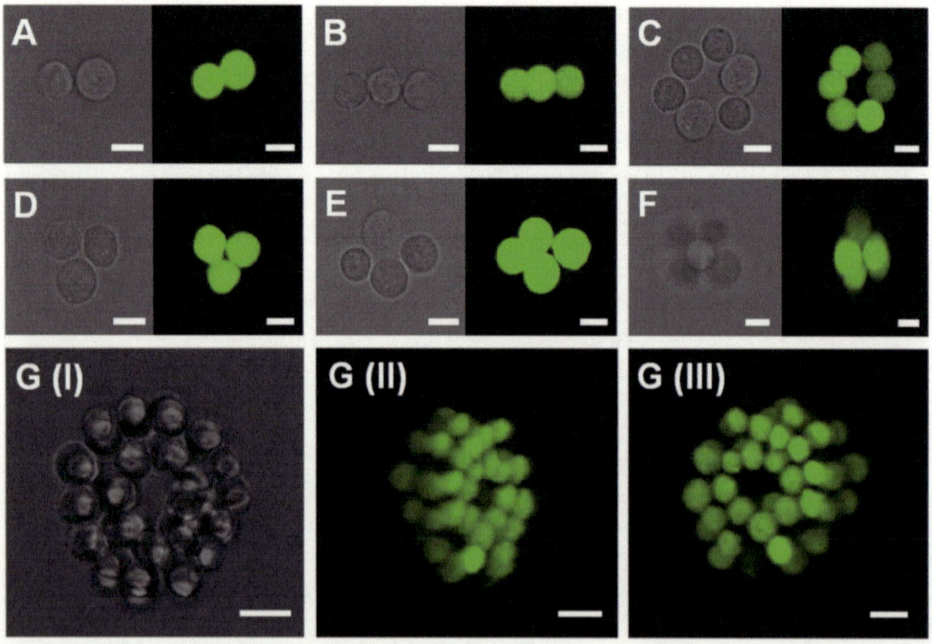

Figure 3.
Bright-field optical micrographs and confocal fluorescence micrographs of one-, two-, and three-dimensional microarray structures of embryonic stem cells assembled by holographic optical tweezers [16].

fiber-based optical tweezers, the distribution of the exiting light field depends on the shape of the fiber tip, which is a highly focused beam, to create a three-dimensional trapping potential. Currently, the tip of the fiber-based optical tweezers is generally designed as a parabolic, spherical, or conical structure. Different shapes of fiber tip can be prepared by physical polishing, heating stretching, chemical etching, and femtosecond laser processing. By changing the physical parameters of the preparation method, such as temperature, speed, time, etc., the shape and size of the fiber tip can be controlled to achieve different functions. **Figure 5b** shows the output light field distribution of a typical tapered fiber. It can be seen that the light is concentrated at the front end of fiber so that the cells can be trapped on the axis of the front end of the fiber and arranged into an ordered structure, as shown in **Figure 5c** [17].

1.3.2 Application of fiber-based optical tweezers

Since the fiber-based optical tweezers have the advantages of simple fabrication, flexible operation, compact structure, and easy integration, it has applications in many fields. For example, Xin et al. used a flame heating and melting taper to prepare a fiber-based optical tweezers with a tapered tip, which enables the capture of submicron-sized polystyrene particles and *E. coli* cells [20, 21]. Xu et al. realized the rotation of single silver nanowires using two tapered fibers, which provide a controlled and optical method for assembling plasmonic nanostructures [22]. Fiber-based optical tweezers will be developed in the direction of high integration and multifunctionality to adapt to lab-on-a-chip and in vivo requirements. In the future, the fiber-based optical tweezers may integrate multiple functions on a single-fiber probe, as shown in **Figure 6**, such as simultaneously capturing, transporting, sorting, stretching, deforming, and rotating various cells and pathogens in the microfluidics or living blood.

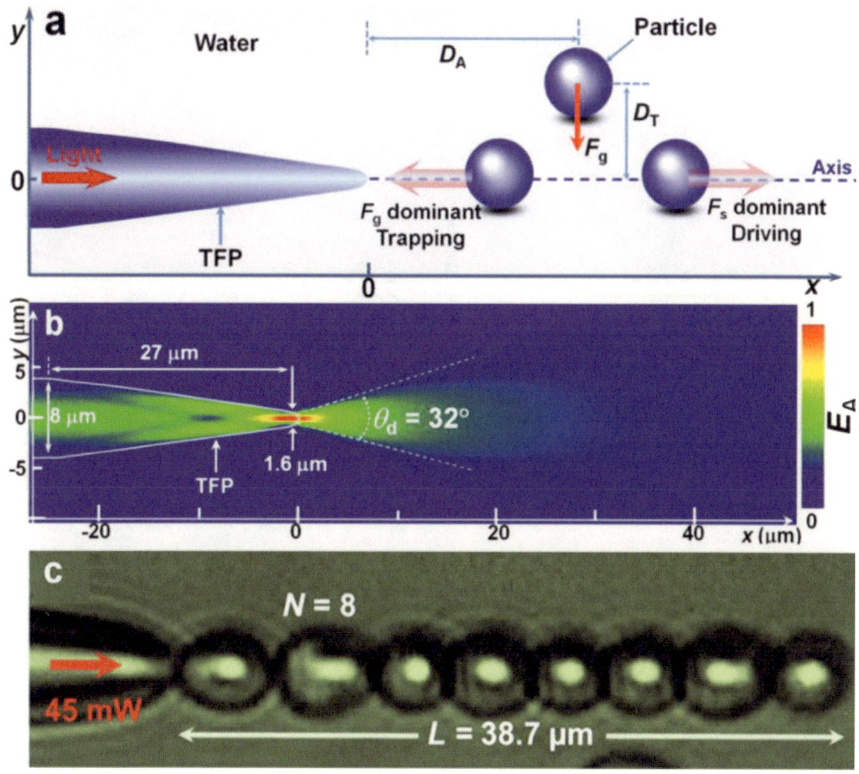

Figure 4.
The basic principle of the fiber-based optical tweezers. (a) Schematic diagram of the optical gradient force (Fg) and scattering force (Fs) applied to the microparticles by the fiber-based optical tweezers. (b) Simulation of electric field intensity distribution of the fiber-based optical tweezers. (c) A chain of yeast cells was trapped by the fiber-based optical tweezers [17].

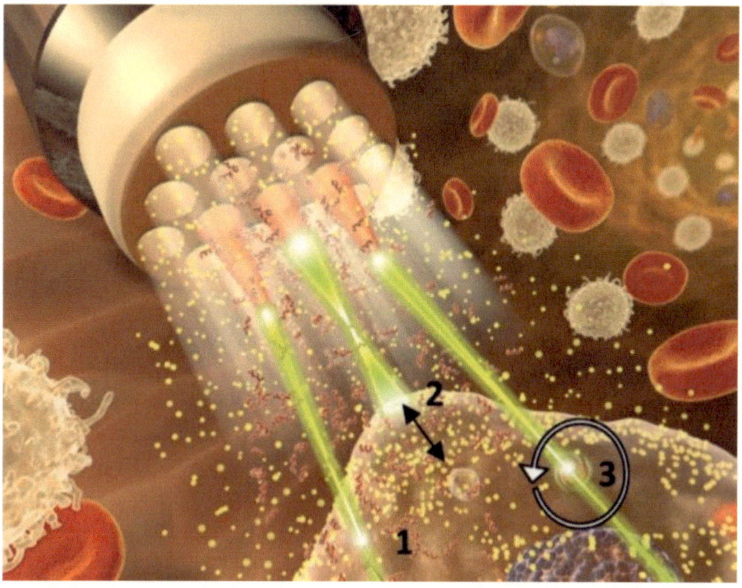

Figure 5.
This schematic shows a versatile fiber-based optical tweezers: number 1 indicates the capture, transport, and sorting of cells, number 2 indicates the optical stretching and deformation of cells, and numbers 3 indicates the optical rotation of cells.

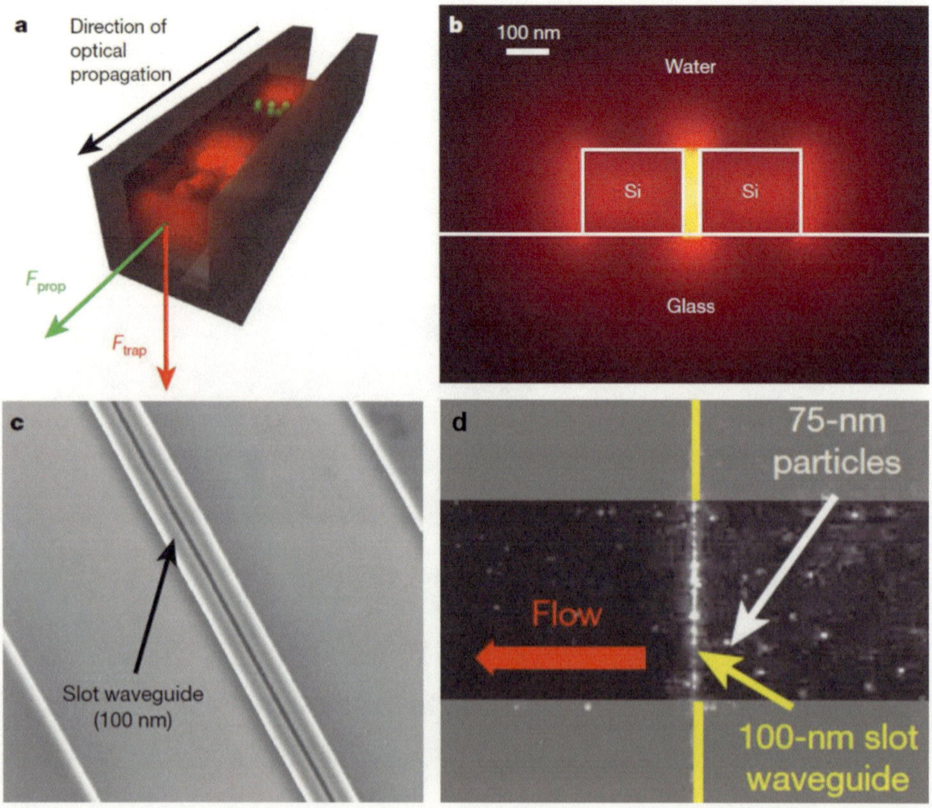

Figure 6.
Slot waveguide optical tweezers. (a) Schematic diagram of the optical gradient force and scattering force of nanoparticles in the slot waveguide. (b) A simulation result of the light intensity distribution of the slot waveguide in an aqueous environment. (c) An electron scanning micrograph of a waveguide having a slit of 100 nm. (d) The slot waveguide captures a large number of polystyrene particles of 75 nm in diameter in the water flow [23].

1.4 Nano-optical tweezers

1.4.1 Planar waveguide optical tweezers

When the light is transmitted in the waveguide, an evanescent wave is generated on the surface of the waveguide due to the total reflection. The evanescent wave is limited to a near-field range of 100 nanometers from the surface of the waveguide. When a nanoparticle enters the evanescent wave, the gradient of the light intensity changes greatly in the direction perpendicular to the waveguide, so the nanoparticles will be trapped on the surface of the waveguide by a strong optical gradient force. In the direction of light propagation, the evanescent wave can be considered to be uniformly distributed. Therefore, there is no optical gravity force in this direction. Only the optical scattering force exists. The nanoparticles move along the direction of light propagation due to the optical scattering force. Therefore, planar waveguide optical tweezers are often used for the transport of nanoparticles. Moreover, since the optical waveguide device is easily integrated into the microfluidic chip, the planar waveguide optical tweezers play an important role in the field of microfluidics. Current planar waveguide optical tweezers can be classified into three types: rectangular waveguide optical tweezers, slot waveguide optical tweezers, and nanofiber waveguide optical tweezers.

The manipulation of microparticles by a rectangular waveguide optical tweezers was first implemented by Kawata et al. [5]. They use rectangular waveguides to perform noncontact optical transport of different sizes of microparticles. This method can deliver cells or drugs over long distances. After this groundbreaking work, more and more researchers have entered this field and designed rectangular waveguides with different structures for transporting metal particles, media particles, microbial cells, etc. [5].

Since the evanescent wave of the rectangular waveguide has limited light confinement, it is challenging for the rectangular waveguide to capture particles and biomolecules below 100 nm. To solve this limitation, the researchers developed slot waveguide nanotweezers [23]. The slot waveguide is an air slit having a width of nanometers by photolithography or electron beam etching. The large refractive index contrast between low refractive index slot and high refractive index waveguide material makes the light energy highly confined in the slot region, which produces a strong optical gradient force and scattering force on the nanoparticles entering the slot. Using this property, Yang et al. achieved capture and transport of polystyrene particles and DNA molecules with sizes below 100 nanometers (as shown in **Figure 7**) [23].

A common problem with rectangular waveguide optical tweezers and slot waveguide optical tweezers is that they must be fixed on the substrate, making it difficult to operate. The emerging nanofiber waveguide optical tweezers can solve this problem. Li et al. used fibers with a diameter of 500–700 nm to achieve stable trap, bidirectional transport, optical separation, and controlled release of nanoparticles and micro-pathogens in microfluidics [26, 27]. The nanofiber waveguide optical tweezers have the advantages of low cost, production, and large control range and have important research value and application prospects in cell transportation, drug delivery, and particle collection.

1.4.2 Photonic crystal optical tweezers

Optical tweezers based on rectangular waveguides, slot waveguides, and nanofiber waveguides can only move particles along the waveguide surface but cannot be used to stably trap nanoparticles. In order to stably capture the nanoparticles, a photonic crystal optical tweezers were developed. The photonic crystal optical tweezers are based on one- or two-dimensional photonic crystal resonator structures (as shown in **Figure 8**) [24, 25]. When the laser that satisfies the wavelength matching condition is coupled into the photonic crystal resonator, static

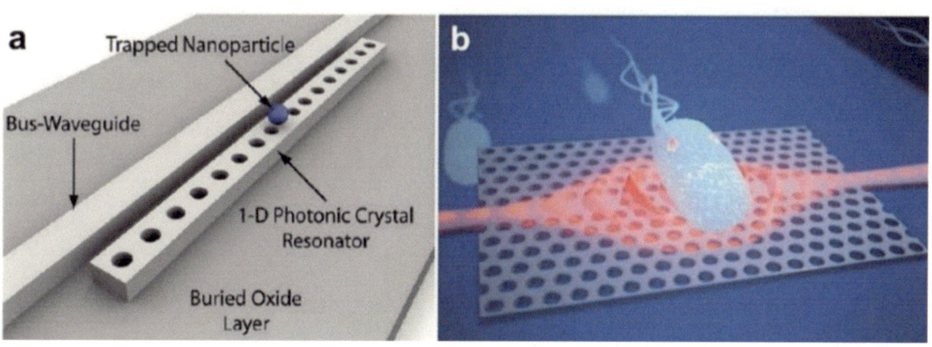

Figure 7.
Photonic crystal optical tweezers. (a) Schematic representation of a single-dimensional photonic crystal resonator capturing a single nanoparticle [24]. (b) Schematic representation of a two-dimensional photonic crystal resonator capturing a single E. coli [25].

interference will occur in the cavity. With the resonance effect, the intensity of the light is greatly enhanced, and the size of the light spot is strongly suppressed, thereby enhancing the optical force of nanoparticles. Based on this principle, Erickson and Mandal et al. achieved stable capture and controlled release of nano-objects such as polystyrene particles, semiconductor quantum dots, and serum protein molecules in a liquid environment [30]. In addition, this method can also be used to study the angular rotation of silver nanowires or carbon nanotubes [31].

1.4.3 Plasmon optical tweezers

Plasmon is a near-field electromagnetic wave formed by the resonance of free electrons on a metal surface and incident photons. Under such resonance conditions, the energy of the electromagnetic field will be converted into the collective vibrational energy of the free electrons on the metal surface, thereby forming a special electromagnetic field: the light is confined to the sub-wavelength of the

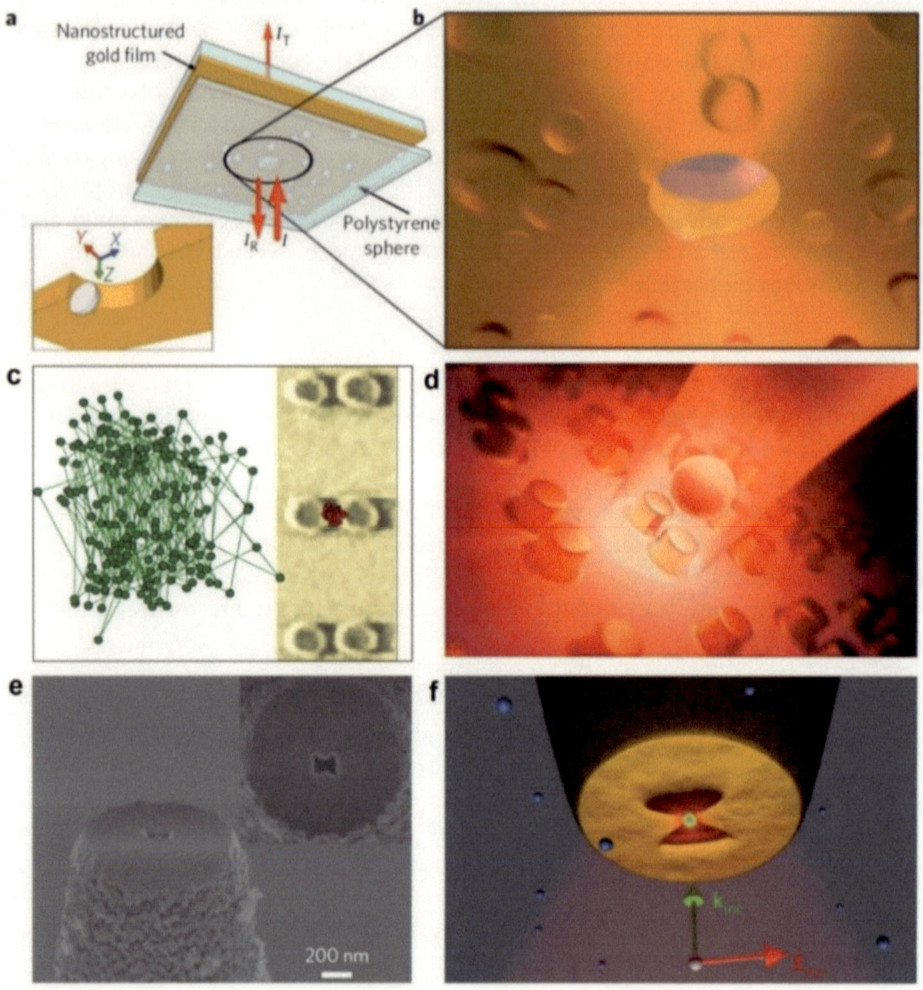

Figure 8.
Plasmon optical tweezers. (a) Schematic diagram of a metal film having nanopores. (b) Schematic diagram of magnified metal nanopore capture nanoparticles. (c) SEM image of the metal nano-antenna structure and motion trajectory after the nanoparticles are captured. (d) Schematic representation of metal nano-antenna structures [28]. (e) SEM image of a metal nano-bowtie structure. (f) Schematic diagram of metal bow nanostructure [29].

metal surface and greatly enhanced. The effect is called the plasmon effect. Since the plasmon effect localizes the light in the near-field range of the nanometer order, it is widely used in the fields of fluorescence signal enhancement, near-field super-resolution imaging, high-density optical storage, integrated optical circuits, etc. [32]. In recent years, the plasmon effect has also been applied in the field of optical trapping and manipulation. The plasmon effect is divided into two types: surface plasmon resonance (SPR) and local surface plasmon resonance (LSPR), both of which can be used to enhance optical force. Researchers used a prismatic total internal reflection to couple incident light into a metal micro-disk on the substrate, which will increase the optical force of the particle by two orders of magnitude and realize the capture of the microparticle. However, the SPR-based optical tweezers can only enhance the optical force of the particle in a two-dimensional plane. Therefore, researchers have proposed an LSPR-based nano-optical tweezers to enhance the optical force of the nanoparticle in three dimensions, including metal nanopores (**Figure 8a, b**), metal nano-antennas (**Figure 8c, d**) [28], metal nano-bows (**Figure 8e, f**) [29], and metal nano-double holes [33]. By using these nano-optical tweezers to achieve trapping of various nanoparticles, such as polystyrene particles, protein molecules, gold particles, micro-pathogenic bacteria, and so on.

2. Conclusions

The noncontact and noninvasive optical trapping and manipulation of microparticles, cells, and biomolecules in liquid environments has broad application prospect in the fields of biomedicine and nanomaterial science [34–47]. Traditional optical tweezers and holographic optical tweezers play an important role in the study of microscale optical manipulation. However, in the rapid development of nanoscience, traditional optical tweezers and holographic optical tweezers are difficult to adapt integration and nano-precision requirements due to the large volume and diffraction limitations. The developed nano-optical manipulation techniques, such as planar waveguides, plasmon optical tweezers, and photonic crystal resonators, can overcome the problem of difficult integration and diffraction limitations of conventional optical tweezers and holographic optical tweezers, which hold great promise in biophotonic and biomedical applications.

Acknowledgements

This work was supported by the National Natural Science Foundation of China (No. 11774135, 11874183, and 61827822).

Conflict of interest

The authors declare no competing financial interests.

Author details

Zhiyong Gong and Yuchao Li*
Institute of Nanophotonics, Jinan University, Guangzhou, China

*Address all correspondence to: liyuchao@jnu.edu.cn

References

[1] Maxwell JC. A Treatise on Electricity and Magnetism. Vol. II. Oxford: Clarendon Press. p. 1873. DOI: 10.1017/CBO9780511709340

[2] Ashkin A. Acceleration and trapping of particles by radiation pressure. Physical Review Letters. 1970;**24**:156. DOI: 10.1103/PhysRevLett.24.156

[3] Ashkin A, Dziedzic JM. Optical trapping and manipulation of viruses and bacteria. Science. 1987;**235**:1517-1520. DOI: 10.1126/science.3547653

[4] Grier DG. A revolution in optical manipulation. Nature. 2003;**424**:810-816. DOI: 10.1038/nature01935

[5] Schmidt BS, Yang AHJ, Erickson D, Lipson M. Optofluidic trapping and transport on solid core waveguides within a microfluidic device. Optics Express. 2007;**15**:14322-14334. DOI: 10.1364/OE.15.014322

[6] Draine BT. The discrete-dipole approximation and its application to interstellar graphite grains. The Astrophysical Journal. 1988;**333**:848-872. DOI: 10.1086/166795

[7] Harada Y, Asakura T. Radiation forces on a dielectric sphere in the Rayleigh scattering regime. Optics Communication. 1996;**124**:529-541. DOI: 10.1016/0030-4018(95)00753-9

[8] Bendix PM, Jauffred L, Norregaard K, Oddershede LB. Optical trapping of nanoparticles and quantum dots. IEEE Journal of Selected Topics in Quantum Electronics. 2014;**20**:15-26. DOI: 10.1109/JSTQE.2013.2287094

[9] Yuan L, Liu Z, Yang J, Guan C. Twin-core fiber optical tweezers. Optics Express. 2008;**16**:4559-4566. DOI: 10.1364/OE.16.004559

[10] Fazal FM, Block SM. Optical tweezers study life under tension. Nature Photonics. 2011;**5**:318-321. DOI: 10.1038/nphoton.2011.100

[11] Padgett M, Bowman R. Tweezers with a twist. Nature Photonics. 2011;**5**:343-348. DOI: 10.1038/nphoton.2011.81

[12] Ashok PC, Dholakia K. Optical trapping for analytical biotechnology. Current Opinion in Biotechnology. 2012;**23**:16-21. DOI: 10.1016/j.copbio.2011.11.011

[13] Heller I, Sitters G, Broekmans OD, Farge G, Menges C, Wende W, et al. STED nanoscopy combined with optical tweezers reveals protein dynamics on densely covered DNA. Nature Methods. 2013;**10**:910-916. DOI: 10.1038/nmeth.2599

[14] Galajda P, Ormos P. Complex micromachines produced and driven by light. Applied Physics Letters. 2001;**78**:249-251. DOI: 10.1063/1.1339258

[15] Dufresne ER, Grier DG. Optical tweezer arrays and optical substrates created with diffractive optics. The Review of Scientific Instruments. 1998;**69**:1974-1977. DOI: 10.1063/1.1148883

[16] Kirkham GR et al. Precision assembly of complex cellular microenvironments using holographic optical tweezers. Scientific Reports. 2015;**5**:8577. DOI: 10.1038/srep08577

[17] Xin H, Li Y, Li B. Controllable patterning of different cells via optical assembly of 1D periodic cell structures. Advanced Functional Materials. 2015;**25**:2816-2823. DOI: 10.1002/adfm.201500287

[18] Dienerowitz M, Mazilu M, Reece PJ, Krauss TF, Dholakia K. Optical vortex trap for resonant confinement of

metal nanoparticles. Optics Express. 2008;**16**:4991-4999. DOI: 10.1364/OE.16.004991

[19] Decombe J-B, Valdivia-Valero FJ, Dantelle G, Leménager G, Gacoin T, des Francs GC, et al. Luminescent nanoparticle trapping with far-field optical fiber-tip tweezers. Nanoscale. 2016;**8**:5334-5342. DOI: 10.1039/C5NR07727C

[20] Xin HB, Li YY, Li LS, Xu R, Li JB. Optofluidic manipulation of Escherichia coli in amicrofluidic channel using an abruptly tapered optical fiber. Applied Physics Letters. 2013;**103**:033703. DOI: 10.1063/1.4813905

[21] Xin HB, Li YY, Liu XS, Li BJ. Escherichia coli-based biophotonic waveguides. Nano Letters. 2013;**13**:3408-3413. DOI: 10.1021/nl401870d

[22] Xu X, Cheng C, Xin H, Lei H, Li B. Controllable orientation of single silver nanowire using two fiber probes. Scientific Reports. 2014;**4**:3989. DOI: 10.1038/srep03989

[23] Yang AH et al. Optical manipulation of nanoparticles and biomolecules in sub-wavelength slot waveguides. Nature. 2009;**457**:71-75. DOI: 10.1038/nature07593

[24] Mandal S, Serey X, Erickson D. Nanomanipulation using silicon photonic crystal resonators. Nano Letters. 2009;**10**:99-104. DOI: 10.1021/nl9029225

[25] van Leest T, Caro J. Cavity-enhanced optical trapping of bacteria using a silicon photonic crystal. Lab on a Chip. 2013;**13**:4358-4365. DOI: 10.1039/C3LC50879J

[26] Xu L, Li Y, Li B. Size-dependent trapping and delivery of submicro-spheres using a submicrofibre. New Journal of Physics. 2012;**14**:033020. DOI: 10.1088/1367-2630/14/3/033020

[27] Li Y, Xu L, Li B. Optical delivery of nanospheres using arbitrary bending nanofibers. Journal of Nanoparticle Research. 2012;**14**:1-7. DOI: 10.1007/s11051-012-0799-3

[28] Grigorenko AN, Roberts NW, Dickinson MR, Zhang Y. Nanometric optical tweezers based on nanostructured substrates. Nature Photonics. 2008;**2**:365-370. DOI: 10.1038/nphoton.2008.78

[29] Berthelot J, Aćimović SS, Juan ML, Kreuzer MP, Renger J, Quidant R. Three-dimensional manipulation with scanning near-field optical nanotweezers. Nature Nanotechnology. 2014;**9**:295-299. DOI: 10.1038/nnano.2014.24

[30] Chen Y-F, Serey X, Sarkar R, Chen P, Erickson D. Controlled photonic manipulation of proteins and other nanomaterials. Nano Letters. 2012;**12**:1633-1637. DOI: 10.1021/nl204561r

[31] Kang P, Serey X, Chen Y-F, Erickson D. Angular orientation of nanorods using nanophotonic tweezers. Nano Letters. 2012;**12**:6400-6407. DOI: 10.1021/nl303747n

[32] Barnes WL, Dereux A, Ebbesen TW. Surface plasmon subwavelength optics. Nature. 2003;**424**:824-830. DOI: 10.1038/nature01937

[33] Kotnala A, Gordon R. Quantification of high-efficiency trapping of nanoparticles in a double nanohole optical tweezer. Nano Letters. 2014;**14**:853-856. DOI: 10.1021/nl404233z

[34] Dholakia K, Reece P. Optical micromanipulation takes hold. Nano Today. 2006;**1**:18-27. DOI: 10.1016/S1748-0132(06)70019-6

[35] Neuman KC, Block SM. Optical trapping. The Review of Scientific Instruments. 2004;**75**:2787-2809. DOI: 10.1063/1.1785844

[36] Leibfried D, Blatt R, Monroe C, Wineland D. Quantum dynamics of single trapped ions. Reviews of Modern Physics. 2003;**75**:281-324. DOI: 10.1103/RevModPhys.75.281

[37] Medintz IL, Uyeda HT, Goldman ER, Mattoussi H. Quantum dot bioconjugates for imaging, labelling and sensing. Nature Materials. 2005;**4**: 435-446. DOI: 10.1038/nmat1390

[38] Saha K, Agasti SS, Kim C, Li X, Rotello VM. Gold nanoparticles in chemical and biological sensing. Chemical Reviews. 2012;**112**:2739-2779. DOI: 10.1021/cr2001178

[39] Lei J, Ju H. Signal amplification using functional nanomaterials for biosensing. Chemical Society Reviews. 2012;**41**:2122-2134. DOI: 10.1039/c1cs15274b

[40] Zhang Y, Liu Z, Yang J, Yuan L. A non-contact single optical fiber multi-optical tweezers probe: Design and fabrication. Optics Communication. 2012;**285**:4068-4071. DOI: 10.1016/j.optcom.2012.06.025

[41] Liberale C, Minzioni P, Bragheri F, De Angelis F, Di Fabrizio E, Cristiani I. Miniaturized all-fibre probe for three-dimensional optical trapping and manipulation. Nature Photonics. 2007;**1**:723-727. DOI: 10.1038/nphoton.2007.230

[42] Pang Y, Gordon R. Optical trapping of a single protein. Nano Letters. 2011;**12**:402-406. DOI: 10.1021/nl203719v

[43] Shoji T, Tsuboi Y. Plasmonic optical tweezers toward molecular manipulation: Tailoring plasmonic nanostructure, light source, and resonant trapping. Journal of Physical Chemistry Letters. 2014;**5**:2957-2967. DOI: 10.1021/jz501231h

[44] Yu XC et al. Single nanoparticle detection and sizing using a nanofiber pair in an aqueous environment. Advanced Materials. 2014;**26**:7462-7467. DOI: 10.1002/adma.201402085

[45] Ndukaife JC, Kildishev AV, Nnanna AGA, Shalaev VM, Wereley ST, Boltasseva A. Long-range and rapid transport of individual nano-objects by a hybrid electrothermoplasmonic nanotweezer. Nature Nanotechnology. 2016;**11**:53-59. DOI: 10.1038/nnano.2015.248

[46] Dholakia K, Reece P, Gu M. Optical micromanipulation. Chemical Society Reviews. 2008;**37**:42-55. DOI: 10.1039/B512471A

[47] Juan ML, Righini M, Quidant R. Plasmon nano-optical tweezers. Nature Photonics. 2011;**5**:349-356. DOI: 10.1038/nphoton.2011.56

Chapter 5

Nanolithography by Scanning Probes for Biorecognition

Javier Martinez

Abstract

With the invention of the scanning tunneling microscope (STM) and subsequently with the atomic force microscope (AFM), the human being was able to enter in the nanoscale world. At first, these devices were only used for imaging samples, but with a small modification of its electronics, they can be used for a precise and controlled manipulation of the scanning probe, creating different types of nanolithographed motifs. The development of this type of lithography has allowed the manufacture of nanometric-scale structures that have led spectacular advances in the field of nanotechnology. In this book chapter, we present the most innovative and reliable probe nanolithography techniques. All of them are based on the spatial confinement of a chemical reaction within a nanometric size region of the sample surface. In that way, 2D or even 3D nanostructures can be fabricated. The full potential of probe nanolithography techniques is demonstrated by showing a range of applications such as the controlled deposition of molecules with high precision or nanotransistors that can be used as sensors for biorecognition processes.

Keywords: AFM, nanotechnology, lithography, nanodevices, scanning probe

1. Introduction

Since the 1960s, the size of electronic devices has been reduced through the use of optical, electronic lithography and lately by immersion lithography. All these techniques are very efficient and allow the fabrication of very complex microelectronic devices, but they have a high cost, are not modifiable, are always made on silicon wafers, and are bounded within the standards of clean rooms. Due to these limitations, in the 1990s, new manufacturing techniques began to be developed that could perform nanometric motifs of different materials in different environments and with lower costs: nanoimprint, soft lithography, and scanning probe nanolithography [1–3].

These new nanofabrication techniques have allowed a great variety of structures to be made, and they have also been able to position and manipulate with nanometric precision different organic and inorganic materials. This chapter wants to provide an overview of the most relevant nanolithography techniques using a local probe that will allow the manufacture of a wide variety of different structures and the creation of functional nanoscale devices for biorecognition.

In order to perform this type of lithography, an atomic force microscope (AFM) is needed, which permits us to obtain high-resolution images in air, in liquids, or in vacuum of all types of conductive, semiconductor, and insulating surfaces. The main elements of the AFM microscope are shown in **Figure 1**. In addition to

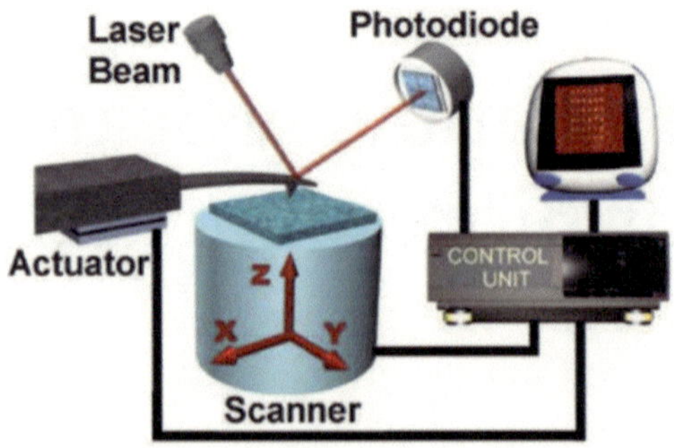

Figure 1.
Main elements of an atomic force microscope system (from Ref. [28]).

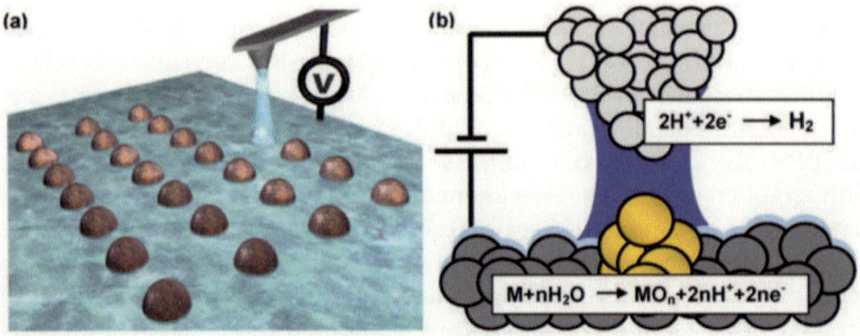

Figure 2.
Local oxidation nanolithography process by AFM. (a) Formation of the liquid meniscus. (b) Chemical reactions in a metallic sample (from Ref. [28]).

generating the image, the control electronics can be configured to use the scanning probe as a powerful tool that can atomically modify the surface with nanometric accuracy. This type of nanometric modifications can be chemical, electrostatic, mechanical, or thermal [4–10].

Nanolithography can be performed under different conditions such as ultrahigh vacuum and low temperature, but the patterning disappears as soon as these conditions are lost, so we will focus on the nanolithography processes at room temperature and without vacuum. In order to perform lithography on a surface, the spatial confinement of a chemical reaction within a nanometric size region is necessary. For this, it is necessary that the probe tip is sufficiently close to the sample so that a liquid meniscus (**Figure 2**) can be formed spontaneous or with the aid of an electric field and also a thermal gradient or a mechanical indentation can be applied.

In a first stage, local oxidation with AFM will be studied in detail. In this case the liquid meniscus that forms between the tip and the sample is water due to relative humidity [10–12]. This type of nanolithography will allow the development of patterns with different shapes [13, 14], and its operating principle can be used to lithograph large areas [15–17]. With this technology, silicon transistors have been made [18, 19]. Afterward it has been observed that these nanotransistors can be used for molecular recognition [20]. In recent years this technique has served to manipulate two-dimensional materials of high scientific interest [21, 22].

By altering the atmosphere where the AFM is housed, nanolithography of different materials can be performed. With octane vapors, extremely small motifs can be obtained [23]. This patterning can be used later for the growth of biological molecules [24]. And by changing to an atmosphere of CO_2, gas molecules can be converted into solid deposits on the surface by applying an electric field [25, 26].

In recent years, a new 3D lithography technique has been developed [27]. In this case, a standard AFM probe has been replaced by a thermal one that reaches a high temperature at its final tip, and the polymer that is deposited on the surface is thermally moldable in three dimensions by scanning with this thermal probe.

2. Local anodic oxidation

Local oxidation of semiconductor, metallic, and organic surfaces by atomic force microscopy (AFM) has established itself as a robust, reliable, and flexible lithographic method for the fabrication of nanometer-scale structures and devices [28, 29].

The invention of this technique appeared in 1990, when Dagata and his collaborators realized that by applying a voltage between the tip of an STM and a silicon sample, their surface was modified and they were able to demonstrate that it was an oxide by mass spectroscopy [12]. A few years later, in 1993 it was done through AFM [30].

The application of a voltage pulse between the tip and the sample polarizes the water molecules in the gas phase and those absorbed on the sample surface. When the voltage is above a certain threshold value, a field-induced liquid meniscus is formed between the tip and sample surface (**Figure 2a**). The water meniscus provides both the chemical species (**Figure 2b**) and the spatial confinement for the anodic oxidation of a nanometer size region of the sample surface [28, 31]. The AFM tip is used as a cathode, and the water meniscus provides the electrolyte.

The size of the oxide motifs can be modified by applying different values of the voltage pulse since it depends linearly. In this way, structures of less than 10 nm have been made reproducible. The voltage pulses are generally between 10 and 30 V and the duration a few milliseconds. The heights of the oxides are a few nanometers, and only 60% of the oxide is above the surface of the sample; the rest is buried in the silicon sample.

This nanofabrication technique allows to perform all types of patterning as can be seen in **Figure 3**: arrays of points, circles, or even the first lines of *Don Quixote* [28].

The process is rather general because many different materials have been patterned such as semiconductors [32], metals [33], dielectrics [34], perovskite oxides [35], or self-assembled monolayers [36].

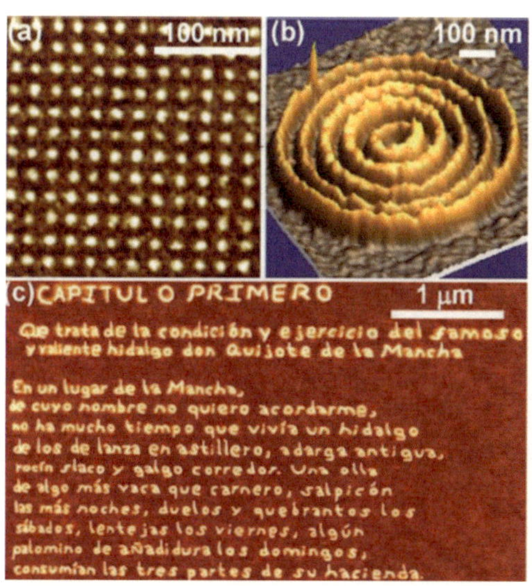

Figure 3.
Examples of local oxidation nanopatterns (from Ref. [28]).

Although nanometric patterns can be generated quite accurately, the main disadvantage of this technology is that AFM is a slow technique and can only cover small areas of a few square microns. To scale this process, a nanoimprint stamp has been developed with millions of protrusions similar to the AFM probe. The stamp has been metallized in order to apply an electric field that allows the oxidation process (**Figure 4a**). The areas of square centimeters with nanometric patterning can be oxidized [16, 17]. An example of that oxidation can be shown in **Figure 4b**, the area is only 5 × 5 μm due to the scan of the AFM, but the oxide patterns are in the whole sample of 1 × 1 cm.

In many of the cases, during the nanofabrication different charges are trapped inside the oxide lines, and that can be used for the selective positioning of molecules [37]. As an example, in **Figure 5** one can observe a controlled deposition of ferritin molecules on the oxide lines made by AFM. For a better positioning, it is necessary to deposit on the silicon sample a self-assembled monolayer of octadecyltrichlorosilane (OTS) and to deposit a monolayer of aminopropyltriethoxysilane (APTES) after the local oxidation [38].

Figure 4.
(a) Scheme of parallel oxidation lithography process with a nanoimprint stamp. (b) AFM image of the silicon oxide line pattern (from Ref. [17]).

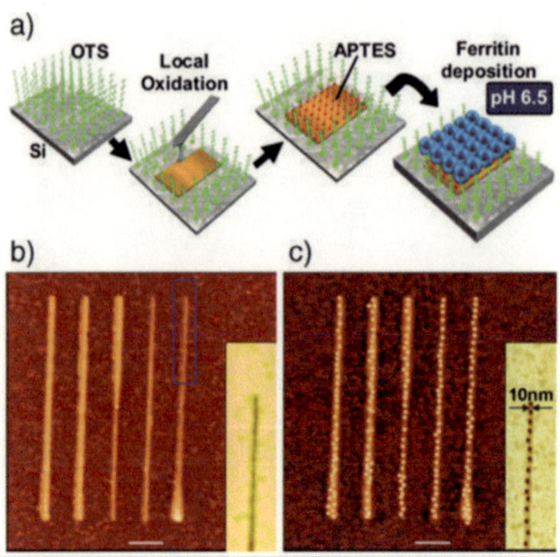

Figure 5.
Patterning of ferritin molecules by local oxidation nanolithography and surface functionalization (from Ref. [38]).

Also with this technology, it is possible to create functional devices. An example, in **Figure 6**, a transistor with a 4 nm silicon nanowire made by local oxidation is shown [18].

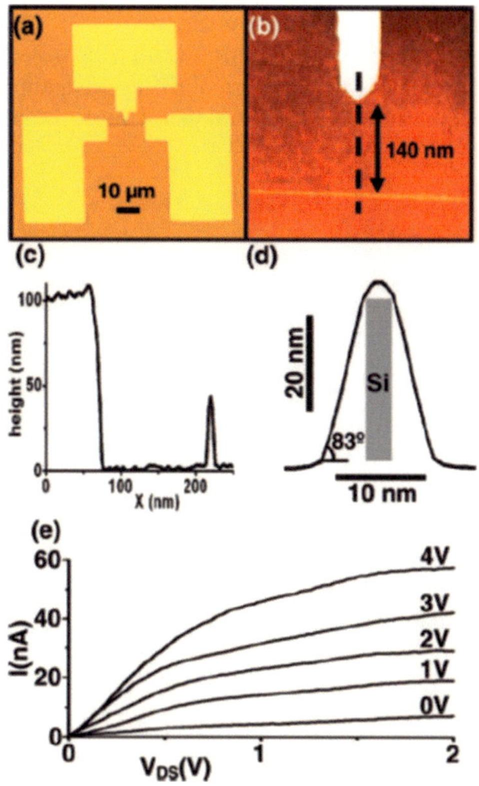

Figure 6.
Silicon nanowire transistor fabricated by local oxidation nanolithography (from Ref. [18]).

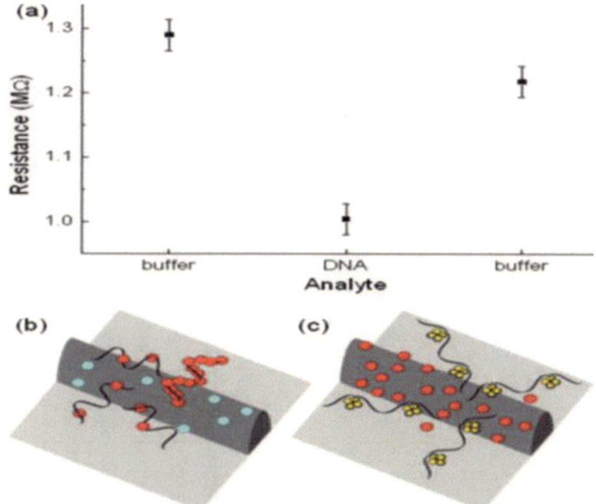

Figure 7.
The silicon nanowire sensor changes its electrical behavior in the presence of DNA and is able to recover its resistance after cleaning (from Ref. [20]).

In this case a silicon on insulator (SOI) wafer was used. The gate, drain, and source contacts were first made by optical lithography. Between these last two, a local oxidation line was made that serves as a mask for the following etching of the top silicon by reactive ions (RIE). In this way the silicon nanowire is free, and after a second stage of lithography and metallization, the source and the drain are in contact with the nanowire forming a nanotransistor.

These nanowire sensors can subsequently be functionalized with different molecules to perform molecular recognition of different agents [19, 20]. In **Figure 7**, a nanowire is used for measuring the early stages of recombinational DNA repair by RecA protein [20].

3. Chemical nanofabrication

Changing the atmosphere surrounding the AFM can produce other chemical reactions between the tip and the samples, which will allow us to manufacture motifs or materials that are not oxides.

For doing this, it is necessary to introduce the AFM into a glove chamber or in a closed environment where it is possible to remove the relative humidity from the environment by a nitrogen flow. Subsequently, the gas to be used for nanolithography is introduced, and an electric field is applied again between the tip and the sample.

Thanks to this type of lithography, polymeric motifs as small as 2 nm resolution at 3 nm at half pitch in ambient conditions have been achieved [23]. This is the smallest periodic pattern fabricated on silicon at atmospheric pressure and room temperature.

The method is based on the formation of a nanoscale octane liquid meniscus between a sharp conductive protrusion and a silicon (100) surface. The application of a high electrical field (10 V/nm) produces the polymerization and cross-linking of the octane molecules within the meniscus followed by their deposition. The manufactured motifs can be seen in **Figure 8**.

This technology can also be used to break up very stable gaseous molecules such as CO_2 and turn them into solid motifs. Thus, if the AFM is introduced into a CO_2

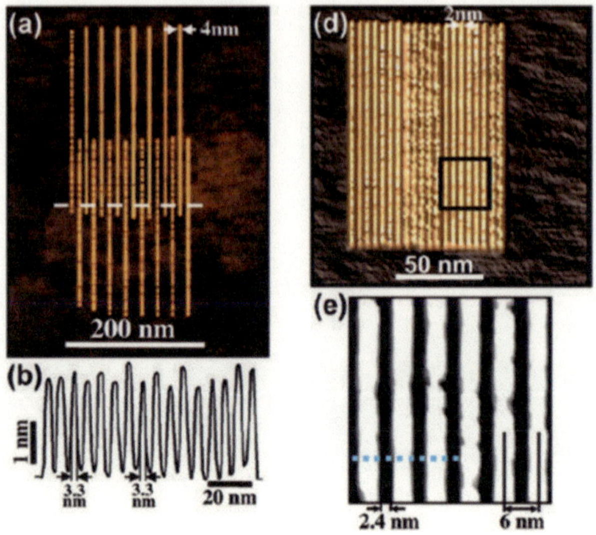

Figure 8.
AFM images and cross sections of the polymeric nanostructures (from Ref. [23]).

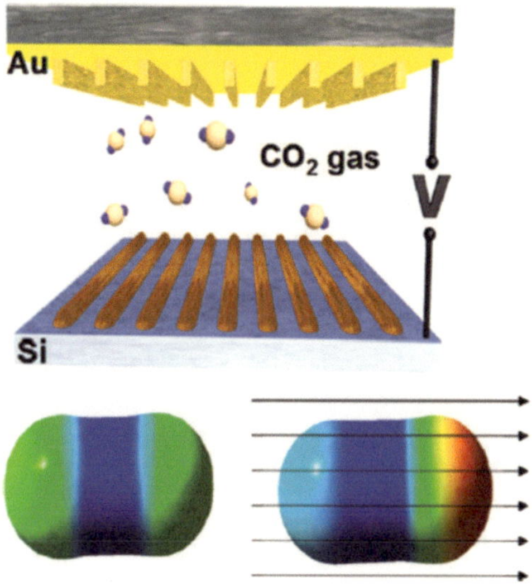

Figure 9.
Conversion of CO_2 gas molecules in solid nanometric motives by applying an electric field (from Ref. [25]).

atmosphere and later an electric field is applied, the CO_2 molecules are able to break due to the high electric field at the end of the tip [25, 26]. This happens for an electric field above 40 V/nm. This technology can be scaled again using PDMS stamp of several square centimeters with thousands of protrusions like in the scheme of **Figure 9**.

The possibilities of generating different nanolithography with different materials are enormous since they only depend on the atmosphere in which the AFM is inserted. The only disadvantage is the need for a spectroscopy analysis after lithography to identify the nature of the motives created.

4. Nanofabrication in 3D

In recent years, micro nanofabrication technologies have advanced quite a bit and are allowing more and more sophisticated AFM tips. Thus, in 2010, IBM laboratories in Zurich made an AFM probes that were doped at their end so that they could behave with a thermal tip when a current is applied [27].

The high temperature at the end of the AFM tip was used to perform a patterning on a glassy organic resist. This local desorption allowed to make structures at a half pitch down to 15 nanometers without proximity corrections. These patterns can be transferred to other substrates, and the material can be removed in successive steps in order to fabricate complex three-dimensional structures (**Figure 10**).

This technique is in continuous development and has great future potential, but it also depends on the thermal tip and on the optimization of the appropriate resins that allow its elimination layer by layer. As can be seen in **Figure 11**, it was possible to make a replica of the Matterhorn mountain in Switzerland first on the resist and then transfer its pattern to silicon.

5. Conclusions

Although the AFM began as a technique to visualize images of a few microns, its potential was seen to be able to manipulate materials in the nanoscale due to various

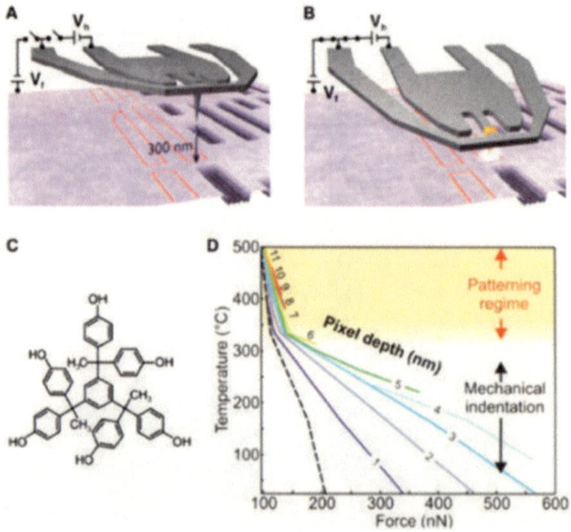

Figure 10.
AFM thermal probe making 3D nanopatterning over a resist (from Ref. [27]).

Figure 11.
(A) AFM picture of the Matterhorn replica in the molecular glass resist. (B) Picture of the Matterhorn mountain. (C) AFM replica in silicon (from Ref. [27]).

reasons. The first of these reasons is the high precision of the piezoelectric devices that allow the AFM tip to be positioned in the right place, and the closed loop control electronics allow a repetitive positioning better than an interferometric stage. On the other hand is the small size of the AFM tip, usually 10 nm or smaller. This allows to obtain liquid menisci of very small volumes in which chemical reactions of various kinds can be created. The small size of the tip also facilitates that with low voltage, high electric fields are obtained at the interface between the tip and the sample, allowing to oxidize different materials or make solid deposits of molecules that are in the vapor phase. Finally in recent years the microelectronic industry has been able to make more sophisticated probes in which they can get the final apex of the tip at very high temperature. This type of tips can modify or even sublimate resins on a surface and can create 3D lithographic motifs that can then be transmitted to the different materials.

With these lithography techniques by scanning probes, great nanotechnological advances have been achieved. The first was to be able to create smaller structures than those achieved by electron beam lithography. It has made possible to lithograph different designs on all types of materials from conductors, semiconductors, or even insulators and more recently in 2D materials like graphene or dichalcogenides.

The second advance was that nanolithographed structures have shown selective positioning of different molecules due to the charges trapped in lithographed motifs. On the other hand, lithographed motifs by scanning microscopy can be used as masks to perform more complex devices such as memories, sensors, or field-effect nanotransistors. These nanotransistors are ideal for its use as sensors for single molecule biorecognition.

In summary, scanning probe nanolithography techniques are very precise and very versatile and constitute an adequate tool for the development of nanotechnology without the need for large and expensive conventional lithography equipment. In addition, the motifs that are capable of manufacturing can be easily scaled for the macroscale simply with the use of nanoimprint techniques.

Acknowledgements

The present research was partially funded by the Spanish Ministry of Science, Innovation and Universities under the project DIGRAFEN, grant number (ENE2017-88065-C2-1-R) and (ENE2017-88065-C2-2-R).

Conflict of interest

The authors declare no conflict of interest.

Author details

Javier Martinez
Institute of Optoelectronic Systems and Microtechnology-ISOM, Technical University of Madrid, Spain

*Address all correspondence to: javier.martinez@upm.es

References

[1] Sotomayor Torres CM. Alternative Lithography: Unleashing the Potential of Nanotechnology. New York: Kluwer Academic/Plenum Publishers; 2003

[2] Geissler M, Xia Y. Patterning: Principles and some new developments. Advanced Materials. 2004;**16**:1249

[3] Quate CF. Scanning probe as a lithography tool for nanostructures. Surface Science. 1997;**386**:259

[4] Binnig G, Rohrer H. In touch with atoms. Reviews of Modern Physics. 1999;**71**:S324

[5] Wouters D, Schubert US. Nanolithography and nanochemistry: Probe-related patterning techniques and chemical modification for nanometer-sized devices. Angewandte Chemie, International Edition. 2004;**43**:2480

[6] Nyffenegger RM, Penner RM. Nanometer-scale surface modification using the scanning probe microscope: Progress since 1991. Chemical Reviews. 1997;**97**:1195

[7] Yeung KL, Yao NJ. Scanning probe microscopy in catalysis. Nanoscience and Nanotechnology. 2004;**4**:1

[8] Cavallini M, Biscarini F, Leon S, Zerbetto F, Bottari G, Leigh DA. Information storage using supramolecular surface patterns. Science. 2003;**299**:531

[9] Sugimoto Y, Abe M, Hirayama S, Oyabu N, Custance O, Morita S. Atom inlays performed at room temperature using atomic force microscopy. Nature Materials. 2005;**4**:156

[10] Samori P. Exploring supramolecular interactions and architectures by scanning force microscopies. Chemical Society Reviews. 2005;**34**:551

[11] Garcia R, Calleja RH. Patterning of silicon surfaces by non-contact atomic force microscopy: Field induced formation of nanometer-size water bridges. Journal of Applied Physics. 1999;**86**:1898

[12] Dagata J, Schneir J, Harary HH, Evans CJ, Postek MT, Bennett J. Modification of hydrogen-passivated silicon by a scanning tunneling microscope operating in air. Applied Physics Letters. 1990;**56**:2001

[13] Tello M, Garcia R. Nano-oxidation of silicon surfaces: Comparison of noncontact and contact AFM methods. Applied Physics Letters. 2001;**79**:424

[14] Ryu YK, Garcia R. Advanced oxidation scanning probe lithography. Nanotechnology. 2017;**28**:142003

[15] Martinez J, Losilla NS, Bisacarini F, Schmidt G, Borzenko T, Molenkamp LW, et al. Development of a parallel local oxidation nanolithography instrument. Review of Scientific Instruments. 2006;**77**:086106

[16] Albonetti C, Martinez J, Losilla NS, Greco P, Cavallini M, Borgatti F, et al. Parallel-local anodic oxidation of silicon surfaces by soft stamps. Nanotechnology. 2008;**19**:434303

[17] Losilla NS, Martinez J, Garcia R. Large área nanoscale patterning of silicon surfaces by parallel local oxidation. Nanotechnology. 2009;**20**:475304

[18] Martinez J, Martinez RV, Garcia R. Silicon nanowire transistors with a cannel width of 4 nm fabricated by atomic force microscope nanolithography. Nano Letters. 2008;**8**:3636-3639

[19] Martinez RV, Martinez J, Garcia R. Silicon nanowire circuits fabricated

by AFM oxidation nanolithography. Nanotechnology. 2010;**21**:245301

[20] Chiesa M, Cardenas PP, Oton F, Martinez J, Mas-Torrent M, Garcia F, et al. Detection of the early stage of recombinational DNA repair by silicon nanowire transistors. Nano Letters. 2012;**12**:1275-1281

[21] Dago AI, Ryu YK, Garcia R. Sub-20 nm patterning of thin layer WSe2 by scanning probe lithography. Applied Physics Letters. 2016;**109**:163103

[22] Dago AI, Sangiao S, Fernández-Pacheco R, De Teresa JM, Garcia R. Chemical and structural analysis of Sub-20 nm Graphene patterns generated by scanning probe lithography. Carbon. 2018;**129**:281-285

[23] Martinez RV, Losilla NS, Martinez J, García R. Patterning polymeric structures with 2 nm resolution at 3 nm half pitch in ambient conditions. Nano Letters. 2007;**7**:1846-1850

[24] Martinez RV, Losilla NS, Martinez J, Tello M, Garcia R. Sequential and parallel patterning by local chemical nanolithography. Nanotechnology. 2007;**18**:084021. DOI: 10.1088/0957-4484/18/8/084021

[25] Calvaresi M, Martinez RV, Losilla NS, Martinez J, Garcia R, Zerbetto F. Splitting CO2 with electric fields: A computational investigation. Journal of Physical Chemistry Letters. 2010;**1**:3256-3260

[26] Garcia R, Losilla NS, Martínez J, Martinez RV, Palomares FJ, Huttel Y, et al. Nanopatterning of carbonaceous structures by field-induced carbon dioxide splitting with a force microscope. Applied Physics Letters. 2010;**96**:143110

[27] Pires D, Hedrick JL, Silva AD, Frommer J, Gotsmann B, Wolf H, et al. Nanoscale three-dimensional patterning of molecular resists by scanning probes. Science. 2010;**328**:732-735

[28] Garcia R, Martinez RV, Martinez J. Nanochemistry and scanning probe nanolithographies. Chemical Society Reviews. 2006;**35**:29-38

[29] Garcia R, Knoll AW, Riedo E. Advanced scanning probe lithography. Nature Nanotechnology. 2014;**9**:577-587

[30] Day HC, Allee DR. Selective area oxidation of silicon with a scanning force microscope. Applied Physics Letters. 2691;**1993**:62

[31] Kuramochi H, Ando K, Tokizaki T, Yokoyama H. In situ detection of faradaic current in probe oxidation using a dynamic force microscope. Applied Physics Letters. 2004;**84**:4005

[32] Mori G, Lazzarino M, Ercolani D, Sorba L, Heuen S, Locatelli A. Desorption dynamics of oxide nanostructures fabricated by local anodic oxidation nanolithography. Journal of Applied Physics. 2005;**97**:114324

[33] Matsumoto K, Gotoh Y, Maeda T, Dagata JA, Harris JS. Room-temperature single-electron memory made by pulse-mode atomic force nanoooxidation process on atomically flat a-substrate. Applied Physics Letters. 2000;**76**:239

[34] Gwo S. Scanning probe oxidation of Si_3N_4 masks for nanoscale lithography, micromachining and selective epitaxial growth on silicon. Journal of Physics and Chemistry of Solids. 2001;**62**:1673

[35] Pellegrino L, Pallecchi I, Marre D, Bellingeri E, Siri AS. Fabrication of submicron-scale $SrTiO_3$ devices by an atomic force microscope. Applied Physics Letters. 2002;**81**:3849

[36] Maoz R, Frydman E, Cohen SR, Sagiv J. Constructive nanolithography:

Inert monolayers as patternable templates for in-situ nanofabrication of metal- semiconductor-organic surface structures. Advanced Materials. 2000;**12**:725

[37] Losilla NS, Oxtoby NS, Martinez J, Garcia F, Garcia R, Mas-Torrent M, et al. Sub-50 nm positioning of organic compounds onto silicon oxide patterns fabricated by local oxidation nanolithography. Nanotechnology. 2008;**19**:455308. DOI: 10.1088/0957-4484/19/45/455308

[38] Martinez RV, Martinez J, Chiesa M, Garcia R, Coronado E, Pinilla-Cienfuegos E, et al. Large-scale nanopatterning of single proteins used as carriers of magnetic nanoparticles. Advanced Materials. 2010;**22**:588-591

Chapter 6

Biophotonics in Africa Powered by Light Technology Applied to Medical Work

Klaudia Freire

Abstract

Biophotonics technologies can be designed to provide unique, dynamic information about tissue structure and biochemical composition. Their impact spans from medical diagnostic and therapeutic devices to consumer-based wearable sensors. With advances in device miniaturization and high-performance biophotonics components, the line between conventional medical instruments and consumer devices is becoming increasingly blurred. Health care economic pressures are further accelerating this ambiguity by shifting clinical attention from expensive disease treatments to strategies for cost-effective disease management and prevention. This clinical research collaboration introduces emerging biophotonics technologies that are capable of characterizing brain tissue structure and biochemical composition spanning from micro to macroscopic regimes.

Keywords: biophotonics, Africa, health technology, light, point of care, medicine

1. Introduction

Biophotonics is a scientific field merging biology and photonics, with photonics being the science and technology of generation, manipulation, and detection of photons, quantum units of light. Research in biophotonics is focused on the development of innovative applications in clinical diagnosis of cancer and related therapies involving fluids, cells, and tissues using light-based tools to excite matter and to transfer information back to the biological operating system.

"Today, healthcare is moving from a treatment-oriented system to a diagnostic oriented one, the end-goal being companioned diagnostic, which is promising but still far away. All these issues have in common a strong need to study real-time evolution of complete living organisms, or part of them (tissues, organs, cells, proteins, DNA, etc.) [1]" and "Therefore, it's essential to develop technologies for quality and process control, as well as rapid microbiological methods suited to the entire production system," says Jacques Cochard, Founder of Tematys.

Global Health Care systems are currently focused on the creation of a well-structured coordination of research effort at global scale aiming to find and implement innovative solutions for sustainable biophotonics devices production and light use as a resource and therefore centered on the need of developing a long term and sustainable partnership in the biophotonics research area that will develop a set of actions devoted to deliver an implementation plan able to stir it and make it effective and operative in terms of sustainable solutions for Health Care systems dealing with new challenges that arise from climate change non described potential epidemics leading to new avenues exploration in terms of scientific research to develop new tools for clinical practice to be applied at point of care in order to improve better clinical protocols outcomes both at low cost and patient patient-centered health care delivery policies.

Diagnosis security is globally recognized as one among the major challenges our society is facing in health care systems. Neurodegenerative diseases are a serious emerging problem in Africa that health care systems need to cope with but the strategies to this novel epidemic ecosystem is far from the optimal resources to ensure the control of epidemical dimension of the problem in Africa. This is particularly the case in rural areas, the most vulnerable regions, exposed to multiple challenges. At the same time, the health care system still plays a key role in improving the life quality in populations by providing support to local rural and urban ecosystems. In fact, health sectors are strategic in the whole African area in terms of future social development, of rural population and territorial development of population in Africa. While natural resources are under climate and population stress, clinical practice must cope with quality requirements imposed by patients and by their ever-changing health patterns. How to apply new diagnostic systems on the population and safeguard their biological machinery health for future quality of life is a great challenge in this area.

Given the scarcity of these biophotonic resources and the increasing neurodegenerative diseases in Africa's population, proper biophotonic research and market output is one of the most crucial issues for the sustainable future of African health care systems. In addition, the challenge of low-cost biophotonic diagnostic devices' scarcity is closely linked to the lack of biophotonic researches, as diseases epidemics in Africa human ecosystem has resulted in an overexploitation of nonscientific based clinical resources and the subsequent risk for population health safety.

This issue requires a more focused approach where the long-term impacts of health care management and biophotonics research use should be considered for ensuring sustainable health care on diseases epidemics provision without harmful effects on the population and the patients in particular. Cooperation in research and innovation is considered of particular importance in order to tackle the most pressing challenges of the biophotonics research area in particular through the development of innovative solutions and the promotion of their adoption for improving the efficiency and sustainability of low-cost biophotonics-based diagnostic devices' production and the safety of clinical practice to treat neurodegenerative diseases in Africa.

2. Africa health care solutions and epidemics overview

Africa is a continent with a complex ecosystem concerning biological entities and specific climate characteristics that enable systematic events of diseases epidemics over population leading to a process phase that we can describe in schematic phases according to World Health Organization monitoring epidemic events assessment identification from level emergence, localized transmission, amplification to reduced transmission being this one the last phase of the process in epidemics. World Health Organization proposal strategic planification to provide response in diseases epidemic events in Africa for the different previous identified phases of diseases epidemics process are first anticipation, containment, mitigation and last eradication.

The standard approach based on emergency plan formula with similar parallel to the territory planning in terms of human demographics criteria often deliver poor cost-effective results as the target for the emergency plan is a group of individuals that are subject to different epigenetics that may present different results on standard protocols applications as current world wide programs apply over African territories for epidemic treatment approach.

According to World Health Organization [2] the response interventions are

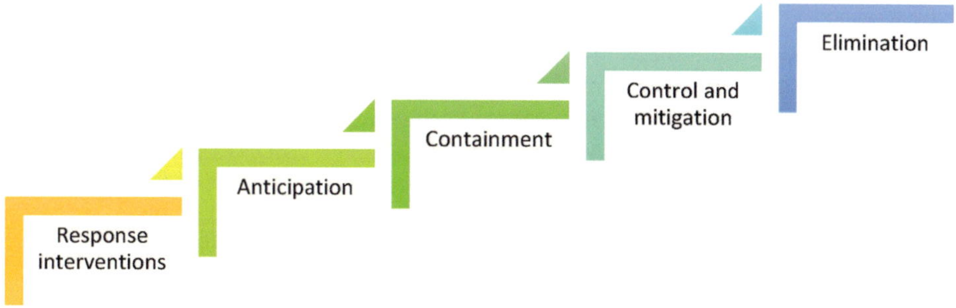

and epidemic phases are

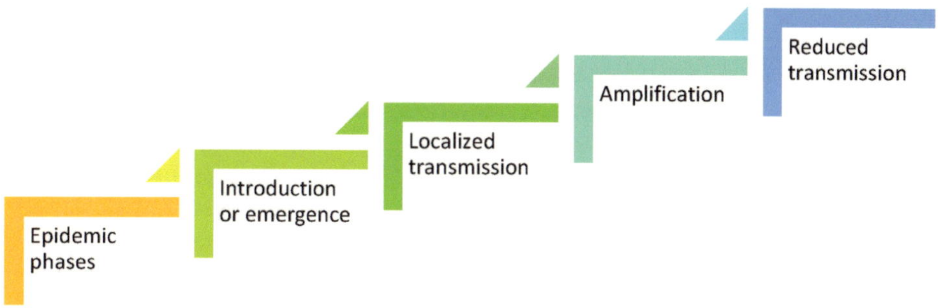

African regions face several epidemic outbreaks that are still a severe threat to national health care systems, which are low-resource settings representing a risk to population health management in the long term.

In order to deal with this critical and systemic situation, African regions receive support from the United Nations World Health Organization and several other global players such as the Gates Foundation Programs, to provide health care solutions for managing epidemics in Africa.

One of the potential solutions for African Governments to accelerate effective response to the epidemics in African regions could be the acquisition of low-cost "top notch equipment," powered by biophotonics technology to enable further impact on the epidemic's detection and eradication.

In recent years, African regions have received significant scientific support concerning biophotonics tools applied to health care with focus on how to "develop cost-effective health care for underserved populations," according to Gerard L. Coté—Director, NSF-ERC on Precise Advanced Technologies and Health Systems for Underserved Populations (PATHS-UP) [3] project development to serve Health Care System on malaria diagnosis in Rwanda with photonics designed technology.

Image: Paths Up Project.

Another example of biophotonics technology applied to health care systems in Africa can be found in Nairobi, Kenya, developed by Prof. Katarina Svanberg [4], leading work in oncology with research work applied to medical lasers to accelerate biomedicine solutions in Africa, empowered by her academic work based at the Department of Oncology in Lund University. Prof. Katarina Shanberg [5] ignited biophotonics education in Senegal. Prof. Katarina Shanberg's group promoted doctor training and offered the medical community photonics-based fluorosensor instruments for trained doctors to perform further biophotonics treatment protocols in Dakar, Senegal, in collaborative work with Sheikh Anta Diop University.

Prof. Katarina Svanberg urges health care systems in Africa to promote the application of biophotonics -based clinical tools at point of care in remote rural areas where patients cannot access health care systems facilities in cities.

The scientific community in Africa is developing one of the firmest research communities dedicated to the exploration of biophotonics applications due to the lack of current medical conventional solutions in African regions, to serve the population suffering from African continent-specific epidemic contexts. This contextual situation is leading to a new generation of African students with academic training in physics and light-based sciences like optics, photonics, and biophotonics searching for solutions in order to accelerate production of light-based clinical instruments to serve African regions in terms of contributing toward health care; according to Zghal [6] "African countries definitely need better-prepared researchers and teachers in optics and photonics to pass on their skills to younger generations."

Africa region is a potential area at global scale that can deliver sustainable biophotonics research and clinical instruments development as Africa needs to accelerate practical low cost solutions to provide innovative Health Care systems with diagnosis and therapeutics for clinical work in African countries, collaborative work has been already ignited with education and training in Africa with the celebration of some events "organized by universities throughout Africa to provide education and training for advanced graduate students and post-doctoral faculty in optics, laser science and technology [7]." like the First African Summer School on Optics and Applications to Sustainable Development [8].

Image: SPIE CEO Eugene Arthurs (at center holding "are" sign) and participants of the Lighting up Africa with Lasers, Optics, and Fibers event in March. In Promoting Optics in Africa https://spie.org/news/spie-professional-magazine/2015-july/optics-in-africa?SSO=1

Africa dares to innovate with biophotonics scientific development in a low resources ecosystem to cope with constant critical epidemic problematics re-emergence that challenge current applied treatments that show less effective solution than biophotonics approach being the critical point for this structural problem in Africa the lack of scientific infrastructures and consolidated educational and training programs in the African Universities and a high level of dependence on educational skills leading to poor scientific results outcomes without external support from other scientific communities participating and contributing to a balanced educational and training environment that is crucial for Africa to mitigate epidemics with a knowledge based own scientific community inside the African territory to increase collaborative research work in the field in a systematic way by monitoring the factors driving to replication and mutation of epidemics outbreaks that requires multidisciplinary scientific and clinical research multidisciplinary approaches for consolidated results.

Investment on biophotonics education and research should be a priority for African Governments to advance with innovative solutions for health care systems "because epidemics are social problems as much as medical ones, we need to move beyond the traditional biomedical approaches to them." as World Health Organization suggests [9], for making progress in emerging health issues in Africa.

Biophotonics represents Africa pipeline focus in sustainable Health Care systems management towards effective results over efforts to eradicate debilitating epidemics in Africa leading to rapid cost-effective growth biophotonics health technology in Africa to deliver smart point of care diagnosis and therapeutic assessments for population health improvement with biophotonics applied to medical protocols. Biophotonics is consequently a significant and promising scientific resource to empower Health Care systems with light based clinical instruments in Africa.

3. Biophotonics for health tech delivery

Biophotonics [10] is the use of light-based technologies in biomedical sciences with a multidisciplinary approach involving the interaction of biology, physics, neuroscience, nanotechnology, and other related fields of scientific knowledge that can provide innovative solutions to products and services development, in this case applied to health care systems.

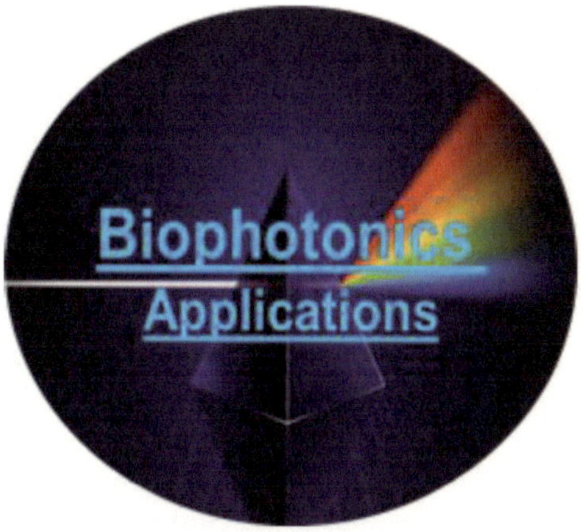

Biophotonics technologies can be designed to provide unique, dynamic information about clinical conditions. The use of biophotonics technology to develop medical diagnostic and therapeutic devices for clinical applications is a growing field of clinical research work worldwide. With advances in devices miniaturization and high-performance biophotonics components, the frontier between conventional medical instruments and innovative light-based clinical application devices is becoming a challenge to doctors that are adapting their own professional skills and resources into a more effective operational clinical work with biophotonics tools.

This issue requires a more holistic approach where the long-term impacts of health care management and biophotonics research use should be jointly considered for ensuring sustainable health care on diseases provision without harmful effects on the population and the patients in particular. Cooperation in research and innovation is considered of particular importance in order to tackle the most pressing challenges in the field of biophotonics in particular through the development of innovative solutions and the promotion of their adoption for improving the efficiency and sustainability of low-cost biophotonics-based diagnostic devices' production and the safety of clinical practices to treat diseases at a global scale.

Biophotonics educational and scientific development practices and policies, through action and investigative research activities, are urgent to promote educational and scientific improvement, and clinical practice resulting in optimal development and high levels of achievement and accomplishments for biophotonics as a leading health technology resource that must be available in the market to health care providers in order to accelerate the development of health innovative products and services with biophotonics technology and related fields of knowledge to serve patients population at global scale because to health care systems, search for information is perhaps the most important problem to minimize the clinical assessment error. In addition, in health care systems, frequently, much time is lost in the search for correct information: this results in downtime and patient's mortality.

Biophotonics systems applied to health care offer the point of integration between the different health care systems and the health care provider operator himself, who performs operations and is the only one who can make decisions and take the actions needed in a very short time, thus reducing costs involved in health care assistance to patient with high quality, which requires smart solutions but simple and inexpensive to simplify clinical protocols at Point Of Care (POC) activities of the assembly operators, in order to reduce errors and increase health care systems' efficiency.

Main economic benefits of biophotonics applied to health

The main benefits for health care providers implementing biophotonics technology systems are:

a. increase in clinical performance because they can get the same health care services delivery at a much lower cost;

b. greater competitiveness in the market;

c. reduction in response times to patients;

d. simplification of clinical protocols activities by health care providers and decision-making processes.

The main end users of biophotonics technology applied to health care are health care provider companies, mainly hospitals, clinics, government NHCS, who often seek to invest in new technologies and improve their health care system services' processes, from older technological systems in departments to flexible and efficient services delivered at point of care with focus on ITC technologies integrations like e-health, telemedicine, and other clinical protocols.

This solution could be particularly attractive for health care providers operating in the market given the expected significant savings in terms of costs.

Time saving+ cost and brain diagnosis time waste reduction+ elimination of the dependence on other systems+ simplification of activities → more efficiency → higher competitiveness

Biophotonics light-based devices for fast diagnosis and theragnostics of low cost and that are intuitive to use are particularly suitable for dynamic realities

a. seeking not too structured systems, flexible to point of care remote clinical needs and

b. wanting fast and simple systems to use and implement.

Rapid demographic, socioeconomic, and climate change factors are threatening the sustainable development of global societies where health care systems must be able to cope with increased demand for innovative diagnostic instruments' production in a scenario of non-invasive diagnostic devices' scarcity in the health care market.

- distributing to the various e-health workstations the ability to decide,
- simplifying and speeding up the decision-making activities in clinical protocols,
- streamlining the activities to be fulfilled by the health care systems elements/operators.

To perform the evaluation and analysis of biophotonics-based innovative solutions from the lab to the marketplace from a technical point of view, several stages of innovation are necessary during systems validation; therefore, technical feasibility studies must be focused on details such as:

- Redesign of the production system
- Device control test
- Detailed energy consumption analysis
- Study on the integration of devices with different production systems
- Control of the information flow
- Compliance with regulatory and safety requirements.

Once the technical feasibility studies are completed, several tests will be performed simulating real working conditions.

The solution will have to be further analyzed from a logistic and from an environmental point of view, to evaluate the clinical diagnosis impacts of the systems to be used.

Integrated systems for biophotonics clinical assessment already exist in the market, in particular, devices for point of care and low-cost applications to be used in smartphones applied to pathologies diagnosis that are able to integrate with

- e-health systems
- health care providers

- national governments' management of healthcare systems

Advantages of biophotonics applied to health care innovative solutions:

- Fewer indirect clinical structures
- Less clinical delivery time
- Ability to keep under control the health care system from all points of view with encrypted data collection but with a real-time action control system
- Monitoring of patient health status in direct time and reducing unnecessary clinical resources waste.

The rationale of the innovation with biophotonics approach health technology is the importance given to innovation which finds solutions that can reinforce competitiveness and better health care results for all the elements involved in the health care process with focus on patient centered clinical assessment for excellent clinical outcomes and patients benefits.

The current challenge for Biophotonics science applied to health care market is to be able to disrupt the medical applications conventional market with powerful clinical tools light based as "Biophotonics research is a field with a history of more than 50 years. Giants like Britton Chance were one of the first scientists to realize the potential of using light in medical applications. Until his death at the age of 96 he was a world leader in the field. He very early transformed theoretical science into useful biomedical applications" [5] for a sustainable diagnosis and theragnosis delivery with light based clinical tools to accelerate global health balance with Biophotonics powered by light technology for the benefit of Humanity. The missing link for sustainable democratic health care systems is technology enabled by biophotonics.

4. Research impact of biophotonics in Rwanda

Current stage of innovation development of BioAdd device has been studying the solution since 2016. Following the Technology Readiness Levels classification, the proposed business innovation project is positioned at TRL 3: technology demonstrated in relevant environment.

All the preceding TRLs (from 1 to 3) have been already tackled and successfully overcome during previous research works. In particular, three major milestones have marked the completed phase "from idea to application" enabling the project to enter the next step "from lab to market":

(M1) Preliminary study and research: this phase was conducted internally in VB lab, through the (human and material) resources of the company. The initial goal was to study how the internal production lines could be improved. Later, an exchange of views with some clients was conducted to verify their needs.

(M2) Construction of an experimental pilot room lab: an experimental system was built (experimental but currently functioning).

(M3) Experimentation successfully carried out on the small pilot room lab, in which some of the technologies here proposed have been tried, obtaining very good feedback from partner to VB Company: IBM Finland.

We can find BioAdd as a diagnosis advanced system applying biophotonics by observing the offered solution:

- Monitoring can take place anywhere and can be done by anyone
- Monitoring can be as frequent as needed with low-cost screening device and App
- Results compare to/beat those obtained by current oncological invasive lab protocols
- Results are available in minutes warning to doctor can be sent within minutes of a test Effectiveness of protocol can be monitored in real time
- Instant analysis and availability of related data improve speed and effectiveness of cancer treatments.

5. Conclusion

Biophotonics represents Africa pipeline focus in sustainable Health Care systems management towards effective results over efforts to eradicate debilitating epidemics in Africa leading to rapid cost-effective growth biophotonics health technology in Africa to deliver smart point of care diagnosis and therapeutic assessments for population health improvement with biophotonics applied to medical protocols. Biophotonics is consequently a significant and promising scientific resource to empower health care systems with light-based clinical instruments in Africa.

Acknowledgements

I wish to express gratitude for the support on the present work by Eng. Ricardo Cabeças, for his critical thinking approach and for his contribution on this work discussion, delivering a proper context view about the innovative technical relevant aspects for biophotonics state of art from biomedical engineering perspective with accurate overview for current scientific development.

Conflict of interest

The author declares no conflict of interest.

Author details

Klaudia Freire
Biomedicum Biophotonics Laboratory, Helsinki, Finland

*Address all correspondence to: claudiaf.univ@gmail.com

References

[1] New technologies will expand the biophotonics market to $36B by 2017 "Biophotonics Market, Focus on Life Sciences & Health Applications" report from Yole Développement & Tematys. http://www.yole.fr/iso_upload/News/2013/PR_BIOPHOTONICS_YOLE%20DEVELOPPEMENT_April2013.pdf

[2] Managing epidemics: key facts about major deadly diseases World Health OrganizationISBN 978-92-4-156553-0 accessed online 05.10.2019 file:///C:/Users/a/Downloads/9789241565530-eng.pdf

[3] Health Systems for Underserved Populations (PATHS-UP). Available from: https://today.tamu.edu/2017/09/12/texas-am-system-to-lead-new-nsf-engineering-research-center/ [Accessed: 05 October 2019]

[4] Katarina Svanberg MD. Professor Lund University Department of Oncology. Available from: https://spie.org/membership/student-services/visiting-lecturers-program/lecturer-directory/svanberg-katarina [Accessed: 05 October 2019]

[5] Sterenborg HJCM, Bendsöe N, Svanberg K. Clinical Biophotonics. Journal of Biophotonics. 2011;**4**(10):665-666. https://doi.org/10.1002/jbio.201100510

[6] Promoting Optics in Africa Community leaders in Africa are increasing the visibility of optics and photonics. Available from: https://spie.org/news/spie-professional-magazine/2015-july/optics-in-africa [Accessed: 05 October 2019]

[7] Tunisia hosts first African Summer School on Optics and Applications to Sustainable Development. Available from: https://spie.org/about-spie/press-room/press-releases/african-summer-school-11-5-2013 [Accessed: 05 October 2019]

[8] First African Summer School on Optics and Applications to Sustainable Development. Available from: https://www.osa.org/en-us/meetings/global_calendar/events/2013/first_african_summer_school_on_optics_and_applicat/ [Accessed: 05 October 2019]

[9] Managing Epidemics Part I. Key Facts about Major Deadly Diseases. Available from: file:///C:/Users/a/Downloads/9789241565530-eng.pdf [Accessed: 05 October 2019]

[10] Make Light Work for Medicine. Available from: https://ec.europa.eu/digital-single-market/node/65248 [Accessed: 05 October 2019]